# BIOFUELS - CHALLENGES AND OPPORTUNITIES

Edited by **Mansour Al Qubeissi**

**Biofuels - Challenges and opportunities**
http://dx.doi.org/10.5772/intechopen.76490
Edited by Mansour Al Qubeissi

**Contributors**

Simeon Iliev, Mansour Al Qubeissi, Juraj Jablonický, Daniela Uhrinová, Juraj Tulík, Ján Polerecký, Armando Pérez, Rogelio Ramos, Gisela Montero, Conrado García, Rene Delgado, Alejandro Suastegui, Héctor Campbell, Marcos Coronado, Fernando De Oliveira, Suani Coelho, Fang Liu, Ryan Davis, Eric Monroe, Razi Ahmad, Dr. Mohd Azlan Mohd Ishak, Luis Adriano Santos Do Nascimento, Geraldo Narciso Rocha Filho, Carlos Emmerson Ferreira Da Costa, Luiza Helena De Oliveira Pires, Deborah Terra Oliveira, Alex De Nazaré Oliveira, Isam Janajreh, Manar AlMazrouei, Paramvir Singh, Sr Chauhan, Varun Goel



First published in London, United Kingdom, 2019 by IntechOpen
IntechOpen is the global imprint of INTECHOPEN LIMITED, registered in England and Wales, registration number: 11086078, The Shard, 25th floor, 32 London Bridge Street
London, SE19SG – United Kingdom
Printed in Croatia

British Library Cataloguing-in-Publication Data
A catalogue record for this book is available from the British Library

Additional hard copies can be obtained from orders@intechopen.com

Biofuels - Challenges and opportunities, Edited by Mansour Al Qubeissi
p. cm.
Print ISBN 978-1-78985-535-7
Online ISBN 978-1-78985-536-4

For EU product safety concerns:
IN TECH d.o.o., Prolaz Marije Krucifikse Kozulić 3, 51000 Rijeka, Croatia,
info@intechopen.com or visit our website at intechopen.com.

# Meet the editor

Al Qubeissi is a Senior Lecturer in Thermo-fluids at the University of Coventry, a Chartered Engineer MIMechE, Fellow of the Higher Education Academy, a member of the International Association of Engineers (IAENG) – Society of Mechanical Engineering (ISME), and Editorial board member of several scientific journals. He is experienced in computational thermo-fluid mechanics. His main relevant research experience and interests include development of mathematical models for processes preceding/exceeding the onset of combustion, conjugate heat transfer systems, and alternative fuel characteristics and their impacts on engines. His research efforts are disseminated in about 80 publications of high impact refereed journal articles, conference proceedings and monographs (a full list of publications can be viewed via https://cumahara.coventry.ac.uk/view/blocks.php?profile=1).

# Contents

# Preface

Fossil sources of energy, including crude oil, coal and natural gas, are non-renewable. They will shrink, if not completely deplete, as the world population is growing and demands for more energy resources are increasing. Yet, the accumulated waste and pollution have caused noticeable damage to the environment and humanity. In addition, there is an immediate need for action to respond to the Climate Agreement of Paris (2015) to sustain minimal global warming to a temperature below 2°C.

The political and public domains provide strong support for the implementation of an affordable and carbon-free electricity system, which at the current time sounds utopian. Moving our current fossil-fuel dependent world to an advanced ideal state requires a time-bridge of transition.

Advanced fuels can be formed from sustainable green resources, with fewer emissions and economical barriers if lands are well managed and agriculture is further modernized. There are many opportunities for rural agricultural development, with a reasonable production of biomass. For efficient and sustainable biomass utilization, a zero-waste approach needs to be embedded. However, with the advancement in processing schemes, biomass is envisaged to remain the exclusive resource of green carbon to be exploited. Although transport sectors will rely largely on electricity, biomass derived hydrocarbon fuels will still be required in many sectors (e.g., aviation), because of the continuous need for high energy density fuels. The abovementioned facts have been investigated and widely appreciated by scientists and engineers. This book, entitled 'Biofuels - Challenges and Opportunities', presents a discussion on these issues, emerging solutions and success parameters.

The book is divided into three sections: Production technologies (Section I); Applications and Challenges (Section II); and Impacts (Section III). Within the three sections, distinguished researchers in relevant fields of study from diverse backgrounds of expertise and nationalities have written nine chapters. The editor acknowledges the authors' contributions, their prompt response in accepting his invitation, and timely delivery of their quality write-ups.

**Dr. Mansour Al Qubeissi**
Coventry University
Coventry, UK

Section 1

# Production Technologies

# Chapter 1

# Introductory Chapter: Biofuels - Challenges and Opportunities

Mansour Al Qubeissi

Additional information is available at the end of the chapter

http://dx.doi.org/10.5772/intechopen.84267

## 1. Background

The climate is undoubtedly experiencing several changes with a global warming, mainly attributed to the carbon dioxide ($CO_2$) emission. According to NASA [1] and NOAA [2], $CO_2$ is at its highest in 650,000 years, leading to the warmest decade on record. During recent decades, concerns have arisen regarding climate change, energy security decline, and depletion of hydrocarbon reserves, resulting in a wide interest in renewable alternative fuel resources. Many studies (e.g., [3–7]) have shown that biofuels are generally the most appropriate replacement to the depleting crude oil.[1] They can expand green landscapes, create new economic opportunities, be directly used in standard engines, and they are sustainable and environmentally friendly. There has been a politically motivated claim; however, that biofuel can be harmful with plenty of disadvantages, attributed to some abuse of biomass production and misuse. For example, moving palm oil with fossil fuel-powered trucks and burning peat bogs to prepare biomass can result in significant greenhouse emissions. Also, economic risks are interpreted with financial loss to certain fossil fuel-productive beneficiaries (e.g., the OPEC). Other claims (e.g., [8]) impose that biofuel production competes with food stock although biofuel is not necessarily addible product, or vigorously agricultural-land planted. Such misleading assumptions and debates guided the European Parliament in January 2018 to propose an end to the import and use of palm oil by 2030 [9]. In fact, the EU move to cut imports on palm oil is a suspect of cynical move to protect the EU vegetable oil producers and fossil fuel-reliant industry [10]. There are enough statistics and empirical data to dispute such claims and evaluate the importance of biofuels for future on the expense of the insignificant disadvantages [3, 7]. This book proposes approaches to overcome challenges and achieves some tough targets for alternative fuel production and utilization.

[1]Hereafter, the word "biofuels" refers to biomass fuels, including biodiesel fuels.

Over the past four decades, the data show three distinct periods of food price inflation, and the lowest of which was during 1991–2015 at 2.6%, encompassing the biofuel boom. Comparatively, the inflation rate during 1981–1991 was 3.8%, and the rate during 1973–1981 was 8.3%. There was no noticeable change in the world's average grain and oil price inflation before and after 2000, in reference to the beginning of the biofuel boom. Crude oil prices, however, have had the highest correlation with food price indices. Currently, the main barriers to the spread of biofuel are its relevant cost of production and competition between some types of biofuels and food stock. Therefore, wastes play an important role in reducing these costs and recycling dump materials. Production of biofuels from nonfood biomass has emerged as a sustainable option to tackle the problems associated with growing demand of energy.

## 2. Contribution of this book

In this work, as well as providing the descriptions on emerging biofuel technologies to mimic the above-mentioned risks, the chemical and thermal properties of biofuels are described. For example, methanol and ethanol are studied as possible alternative liquid fuel candidates to resemble several physical and combustion properties of gasoline. The study shows that such a fuel blend decreases the engine brake power, increases the BSFC, and decreases the CO and HC emissions, compared to those produced using gasoline (fossil) fuel. The pretreated samples are analyzed for mass and energy yields, calorific values, proximate and elemental compositions, and thermal decompositions. This book also provides basic analyses of diesel, gasoline, and various types of biofuels (including biodiesel fuels) and assessment on limited and unlimited emissions (e.g., greenhouse gases, dangerous exhaust gases, and strong carcinogens and their contents) during and after combustion pathways. The results are evaluated in comparison to trusted measurements and numerical standards. The research equipment is adapted with sensitivity measurements to the environmental contamination.

A one-dimensional model of gasoline engine is developed for predicting the effect of various fuel types on engine performance, specific fuel consumption, and emissions. The role of microbial consortium-based biocatalyst strategies that are being developed to address these issues are reviewed and discussed. Microbial co-culture biocatalysts are engineered to specialize the conversion of a general class of substrates present in the biomass hydrolysates into biofuel intermediates, with the capability of adapting them to the variable composition of the feedstock. The techniques being developed to understand the interaction between the members of the bioconversion consortia and the corresponding population dynamics of the engineered co-cultures are also presented. The simulation of transesterification requires in-depth understanding of the chemical reactions that take place inside the reactor. The development of reaction mechanism of the multiple step triglyceride, triglycerides, and monoglycerides and their reversal reaction is beyond the interest of chemical or mechanical engineers, whose main interests focus on the assessment of the overall conversion and the established performance process metrics. The work undertakes all relevant activities by establishing and formulating the overall process kinetics, as far as the rate constant and activation energy. The obtained values are used to carry out high fidelity reactive flow of the multiple

http://dx.doi.org/10.5772/intechopen.84267

species, copresent inside the reactor, and otherwise complex to be captured experimentally. Experimental results, high-fidelity numerical results, and parametric sensitivity studies will be introduced and discussed. Through a reaction of alkaline transesterification of biodiesel, several mixtures of diesel-biodiesel and their characteristics are assessed and prepared for tests. For example, kinematic viscosity and high heating value of pure biodiesel (B100), pure diesel, and four biodiesel-diesel fuel blends, B2, B5, B10, and B20, have been determined.[2] B100, diesel, and their blends are used in full-cycle tests of four cylinder engines. A virtual instrumentation technology has been developed and implemented into the test approaches. Such advancement has allowed monitoring (in real time) the parameters of internal combustion engines and presented the versatility, flexibility, scalability, and capacity to function in equipment that operates with different liquid fuels at a lower cost than that of conventional systems. These characteristics represent significant benefit, in comparison to the classical measurement and monitoring approaches, in present market.

## 3. Concluding remark

To conclude, the use of biofuels is expected to contribute to the energy sustainability and reduction of global warming. For instance, many efforts have been made to replace gasoline and diesel fuels with ethanol/gasoline and biodiesel/diesel fuel blends, respectively [11–16]. These efforts have been driven mainly by the importance of reducing greenhouse emissions and fossil fuel costs [17–19]. According to the US environment protection agency [20], all gasoline engine vehicles can use a blend of gasoline fuel with up to 10% volume fraction of ethanol without the need for engine modification. The reduction in $CO_2$ emissions without the loss of engine performance is nontrivial for this mixture [21]. For example, mixtures with up to 15% volume fraction of ethanol and 85% volume fraction of gasoline fuel have been approved for use in 2001 and newer vehicles, under the US federal standards for renewable fuel [22]; while mixtures with up to 85% volume fraction of ethanol and 15% volume fraction of gasoline fuel (i.e., flex fuels) have been defined as qualifying alternative fuels for flex-fuel vehicles [20]. Therefore, it was important to investigate the difference between ethanol and gasoline fuel characteristics and their blends.

## Author details

Mansour Al Qubeissi

Address all correspondence to: ac1028@coventry.ac.uk

Coventry University, Coventry, UK

[2]BX refers to X% volume fraction of biodiesel mixed with (100-X)% volume fraction of diesel fuel.

## References


[1] NASA. Carbon Dioxide, Climate Change – Vital Signs of the Planet [Internet]. Available from: https://climate.nasa.gov/vital-signs/carbon-dioxide/ [Accessed 2018-12-24]

[2] NOAA. November 2018 was fifth hottest on record for the globe 2018 [Internet]. Available from: https://www.noaa.gov/news/november-2018-was-fifth-hottest-on-record-for-globe [Accessed: 2018-12-25]

[3] Popp J, Lakner Z, Harangi-Rákos M, Fári M. The effect of bioenergy expansion: Food, energy, and environment. Renewable and Sustainable Energy Reviews. 2014;**32**:559-578. DOI: 10.1016/j.rser.2014.01.056

[4] Bullis K. A Biofuel Process to Replace All Fossil Fuels. MIT Technology Review 2009 [Internet]. Available from: https://www.technologyreview.com/s/414492/a-biofuel-process-to-replace-all-fossil-fuels/ [Accessed: 2018-12-28]

[5] Sazhin SS, Al Qubeissi M, Kolodnytska R, Elwardany AE, Nasiri R, Heikal MR. Modelling of biodiesel fuel droplet heating and evaporation. Fuel. 2014;**115**:559-572. DOI: 10.1016/j.fuel.2013.07.031

[6] Walker GM. Bioethanol: Science and Technology of Fuel Alcohol. UK: Bookboon; 2010. 114 p. ISBN 9788776816810. Available from: https://bookboon.com/en/bioethanol-science-and-technology-of-fuel-alcohol-ebook [Accessed: 2018-12-12]

[7] Kirkland G. Why We Need Biofuels For A Green Future. Automotive IQ 2018 [Internet]. Available from: https://www.automotive-iq.com/powertrain/articles/why-we-need-biofuels-green-future [Accessed: 2018-12-21]

[8] Tenenbaum DJ. Food vs. fuel: Diversion of crops could cause more hunger. Environmental Health Perspectives. 2008;**116**(6):A254-A257. DOI: 10.1289/ehp.116-a254

[9] Transport & Environment. Why is palm oil biodiesel bad? In: Food-based biofuels: cure worse than the disease [Internet]. June 2017. Available from: https://www.transportenvironment.org/what-we-do/biofuels/why-palm-oil-biodiesel-bad [Accessed: 2018-12-30]

[10] Klepper G. Winners and losers from the EU's proposed ban on palm oil [Internet]. Eco-Business. 2018. Available from: https://www.eco-business.com/opinion/winners-and-losers-from-the-eus-proposed-ban-on-palm-oil/ [Accessed: 2018-11-30]

[11] Al Qubeissi M, Al-Esawi N, Sazhin SS, Ghaleeh M. Ethanol/gasoline droplet heating and evaporation: Effects of fuel blends and ambient conditions. Energy & Fuels. 2018;**32**: 6498-6506. DOI: 10.1021/acs.energyfuels.8b00366

[12] Al Qubeissi M. Predictions of droplet heating and evaporation: An application to biodiesel, diesel, gasoline and blended fuels. Applied Thermal Engineering. 2018;**136**: 260-267. DOI: 10.1016/j.applthermaleng.2018.03.010

[13] Al Qubeissi M. Heating and Evaporation of Multi-Component Fuel Droplets. Stuttgart, Germany: WiSa; 2015

[14] Al Qubeissi M, Sazhin SS, Elwardany AE. Modelling of blended diesel and biodiesel fuel droplet heating and evaporation. Fuel. 2017;**187**:349-355. DOI: 10.1016/j.fuel.2016.09.060

[15] Al Qubeissi M, Al-Esawi N, Kolodnytska R. Atomization of bio-fossil fuel blends. In: Nageswara-Rao M, editor. Advances in Biofuels and Bioenergy, Chapter 4. InTech; 2018. pp. 59-81. DOI: 10.5772/intechopen.73180

[16] Al Qubeissi M, Sazhin SS, Crua C, Turner J, Heikal MR. Modelling of biodiesel fuel droplet heating and evaporation: Effects of fuel composition. Fuel. 2015;**154**:308-318. DOI: 10.1016/j.fuel.2015.03.051

[17] Járvás G, Kontos J, Hancsók J, Dallos A. Modeling ethanol-blended gasoline droplet evaporation using COSMO-RS theory and computation fluid dynamics. International Journal of Heat and Mass Transfer. 2015;**84**:1019-1029. DOI: 10.1016/j.ijheatmasstransfer.2014.12.046

[18] Bader A, Keller P, Hasse C. The influence of non-ideal vapor-liquid equilibrium on the evaporation of ethanol/iso-octane droplets. International Journal of Heat and Mass Transfer. 2013;**64**:547-558. DOI: 10.1016/j.ijheatmasstransfer.2013.04.056

[19] Corsetti S, Miles REH, McDonald C, Belotti Y, Reid JP, Kiefer J, et al. Probing the evaporation dynamics of ethanol/gasoline biofuel blends using single droplet manipulation techniques. The Journal of Physical Chemistry A. 2015;**119**:12797-12804. DOI: 10.1021/acs.jpca.5b10098

[20] EPA U. US Environmental Protection Agency 2014. http://www.epa.gov/ [Accessed: June 23, 2017]

[21] Masum BM, Masjuki HH, Kalam MA, Rizwanul Fattah IM, Palash SM, Abedin MJ. Effect of ethanol-gasoline blend on NOx emission in SI engine. Renewable and Sustainable Energy Reviews. 2013;**24**:209-222. DOI: 10.1016/j.rser.2013.03.046

[22] US Department of Energy: Energy Efficiency and Renewable Energy. Alternative Fuels Data Centre n.d. Available from: http://www.afdc.energy.gov [Accessed: June 23, 2017]

Chapter 2

# Valorization of Wastes for Biodiesel Production: The Brazilian Case

Luís Adriano Santos do Nascimento,
Deborah Terra de Oliveira, Alex Nazaré de Oliveira,
Luiza Helena de Oliveira Pires,
Carlos Emmerson Ferreira da Costa and
Geraldo Narciso da Rocha Filho

Additional information is available at the end of the chapter

http://dx.doi.org/10.5772/intechopen.81879

**Abstract**

This chapter intends to bring an overview about the Brazilian researches and their contributions to the production of biodiesel from wastes. Currently, the main obstacles to spread the use of biodiesel are its high cost of production and the competition between biodiesel and food industries. So, the use of wastes plays an important role in reducing the biodiesel costs and reusing the materials that have no other applications, as deodorization residues, neutralization soap sticks, and animal fats, among others. Then, we present a review about Brazilian studies involving waste oils and fatty–acid-rich raw materials that helped the advancement in this field of knowledge during the last few years.

**Keywords:** waste valorization, Brazil, biofuels, biodiesel

## 1. Introduction

Since the 1950s, Petroleum has become the most important source of raw material for industrial chemistry and of fuel. Although in recent years, fossil fuel reserves have been provided for the next two decades, particularly heavy oil, tar sands, and deepwater drilling and oil and shale gas have grown, there are large uncertainties in the economy of their exploitation through current methodologies. Therefore, in the future, renewable resources should regain their importance, since the choice of the raw material acquires an ever greater significance, not only for economic reasons but also because the initial choice will determine the properties of its derivatives and, consequently, the environmental impacts [1–3].

Non-renewable natural resources, fossil fuels, will not be sustainable in the next century due to the high consumption rate (the rate of global energy use is expected to increase from its current level of about 20–27 TW by 2050 and can reach 43 TW by 2100) [1–3] and the serious environmental problems that accompany their combustion. There is an urgent need to develop new forms of energy that are clean, renewable, and low-cost.

In this sense, several countries have developed treaties proposing a gradual reduction in the use of fossil fuel derivatives, as well as an increase in actions aiming the technological development of biofuel production and consumption, and in this way, to minimize the growing signs of global climate change [4, 5]. Moreover, the interest in the world-wide research in improving the production of better-quality fuels that cause less damage to the environment as well as lower expenses for the producers [6] and Brazil follows the world trends.

Brazil presents a highlighted position about the countries in the world, having 18% of fuel sources and 45% of the energy consumed of renewable while all over the world has 14% of the energy coming from renewable energy sources. In the search for the alternative sources to the use of oil, Brazil presented itself as a pioneer country in the use of biofuels, standing out from many countries that search for sources of renewable energy [7]. For instance, the Brazilian production of alcohol in 2009 reaches 25,866.06 $m^3$, in contrast to 12,588.00 $m^3$, before flex-fuel vehicle implantation. For biodiesel, the Brazilian production (**Figure 1**) was 736 $m^3$ in 2005 to 3,419,838 $m^3$ in 2014 [8].

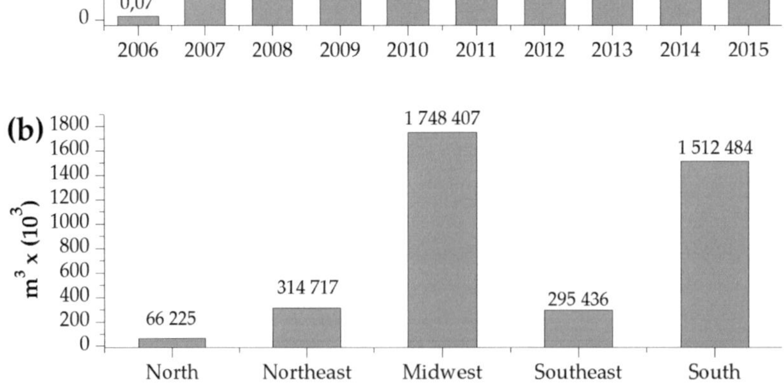


**Figure 1.** (a) Evolution of biodiesel production; (b) production of biodiesel according to large Brazilian regions in 2015 [4].

This highlighted position can be explained for some reasons. First, Brazil is a tropical country with continental dimensions that allow the exploitation of traditional crops such as sugarcane, soybean, peanut, sunflower, castor bean, and palm oil, as well as new alternatives such as tucumã, babaçu, pequi, buriti, jupati, and a wide variety of oilseeds to be explored. Furthermore, due to its great size and variety of climates with many regions of pastures, it is one of the greatest exporters of bovine meat of the world, which can improve the biodiesel production, as can be seen ahead.

Another reason is its experience to produce biofuels. The first reports of the Brazilian experiments with the production of biodiesel and ethanol are from the 1920s, when the National Institute of Technology began to study these biofuels [9, 10]. However, the high consumption of biofuels in the country occurred in the 1970s, with the implementation of the Brazilian biofuel program PROÁLCOOL (National Alcohol Program). With the experience acquired in the implementation of PROÁLCOOL, the Brazilian government launched in December 2004 the National Program for the Production and Use of Biodiesel (PNPB) [10]. In 2005, it was established by law 11,097 that biodiesel should be blended with petroleum oil in 2% (B2), increasing to 5% (B5) in 2013. Under these conditions, the National Petroleum Agency (ANP) is responsible for regulating and complying with established rules for the commercialization of biofuels, which are currently defined as 8% (B8) of biodiesel in petroleum diesel [7].

Currently, Brazil stands out worldwide for the production of sugarcane alcohol and soybean biodiesel, both considered of first generation. However, their current limitation is their high price, which sometimes approaches the cost of fossil diesel fuel. Therefore, the study of alternative raw materials is of major importance for not competing with the food market and reduces production costs, therefore increasing the economic competitiveness of biofuels [11].

In this way, Brazilian researchers have been studying some alternatives to improve the contribution of these biofuels to the energy matrix. In the case of biodiesel we can cite microalgae [12] and cyanobacteria [13] produced naturally, and some types of non-edible plant oils, such as *Jatropha curcas*, castor bean, Murumuru, buriti, and pracaxi, can play a considerable role in the supply of alternative feedstock. On the other hand, they require a large available farmland [7, 14]. Then, a largely accepted strategy to reduce the cost of biodiesel is the use of wastes for the production of biodiesel; besides bringing substantial environmental benefits, it provides an alternative for the final disposal previously discharged in the environment [7].

Based on this context, this chapter intends to bring a review of researches developed by Brazilian researchers in the field of recovering residues, using different types of approaches related to the subject of Biodiesel.

## 2. Waste valorization and the biodiesel in Brazil

There is no doubt that the valorization of wastes plays an important role in the field of biofuel generation due to the low price of such feedstocks, which is a fundamental requirement to reduce prices and to increase the competition of biofuels compared to fossil fuels. However, there are several advantages in reusing materials that could damage the environment. It could reduce the pollutant disposal and could possibility increase the employment generation due to the great amount of people that could be involved in the waste recovery and in the biofuel industry. All these findings enforce researchers to evaluate the feedstocks, the best conditions, and the techniques to produce such waste-based biofuels.

From now on, some literature reports (from the last 15 years) will be presented in this point of view. It is important to point out that there are a great number of scientific articles related to this subject using different approaches. From the study of techniques to characterize the different wastes and waste-based biofuels to the study of purification methods of these biofuels, all are reported. Socioeconomic analysis of the waste valorization and the engine tests are also found in the literature.

Despite such diversity, when focusing in waste valorization, it is possible to point out two wastes that have been studied and whose biodiesels have already been produced (even on a microscale): cooking oil and cattle fat. For this reason, both fats will receive a certain highlight in this manuscript. However, other aspects and examples will be mentioned elsewhere.

### 2.1. Waste cooking oil or frying waste oil (WCO or FWO)

The price of biodiesel is currently the highest selling price since depending on its source it can cost up to 1.5 times more than diesel oil. One way to convey this problem would be the use of oil sources that are the least expensive, such as waste and tailings. The cooking oil used is a raw material that has been used that reduces costs and decreases the amount of material in the environment. In Brazil, the recycling of frying oil is receiving importance and investments from companies interested in carrying out the proper disposal of this waste as well as having an outlook toward profit. Brazil already produces biodiesel from WCO; however, the amount of WCO in the energy matrix is still very low. For instance, just 0.5% of Brazilian biodiesel was produced from WCO in 2015. This value could be very higher, mainly if we consider that Brazil has an estimate of generation of used oils close to 329,020 t/year [7, 15].

Some rural places had problems with de supply of fuel; the WCO would be a good source for biofuels and will solve this problem and besides will change the centralization of the biodiesel production [15].

Some Brazilian researches have been important to increase the knowledge about the WCO biodiesel to improve its role in the energy matrix, as can be seen below:

- In 2000, Costa Neto et al. [16] have presented some comparisons of the quality between the frying oil used as a source of biodiesel and diesel. Were tested your performance in diesel engines even as the characterization of the emissions derived from this process, being observed a significant reduction of fume when biodiesel from waste oil was used. In that year they already alert concerning the importance of a National Program to promote the biodiesel use, which be created just in 2004.
- Hocevar et al. [17] determined the composition in percentage of waste frying oil blends with soy oil, palm oil, and hydrogenated fat by some techniques such as mid-infrared spectroscopy in association with multivariate chemometric analysis. Using this method, a

first estimation of the principal component of oil blends was made possible, as well as the partial least square regression was used to predict the content of each type of oil present in blends of waste frying oil. Then, they showed that this methodology could be very useful for the rapid, reagent-free, low-cost determination of the composition of waste cooking oil and can aid in decision-making regarding the pretreatment of oils and the production routes for biodiesel.

- The current production of biodiesel uses alkaline liquid catalysts that lead to some drawbacks of production like impossibility to reuse the catalyst as well as high corrosiveness of the catalyst. This situation leads to the use of solid catalysts that also show some problems found in the reactive process like the leaching and deactivation of the catalyst in the presence of water during the reaction of esterification of FFA. Considering all this information, in a study of 2011, Silva et al. [18] evaluated the catalytic activity of tin chloride, $SnCl_2$, which is less corrosive, water tolerant, and a recyclable Lewis acid catalyst, in FFA ethanolysis using waste cooking oil samples (WCO) obtaining success reaching high conversions of FFA into FAEE (fatty acid ethyl esters) in almost all catalytic runs; even when compared to pTSA(*p*-toluenesulfonic acid), the SnCl2 catalyst efficiently promoted FFA ethanolysis in the presence of high water concentrations (ca. 5.0% w/w).

- Alberici and co-workers [19] showed that the biodiesel produced from used frying oil show better quality when compared with other common oils. The parameters of quality of biodiesel like the induction period, the acid number, the iodine value, and the heat of combustion were tested in this biofuel. They also reported that an artificial antioxidant, N,N′-di-sec-butyl-p-phenylenediamine, was shown to significantly increase the oxidative stability of the used frying oil biodiesel at a trace level.

- In 2016, Vescovi et al. [20] produced biodiesel from lipase-catalyzed (immobilized *Thermomyces lanuginosus* lipase and immobilized *Candida antarctica* lipase B) hydrolysis of waste cooking oil (WCO) followed by esterification of the hydrolyzed WCO (HWCO). The hydrolysis of acylglycerols was almost complete after 12 h (ca. 94%), and in the esterification step, the conversion was around 90% after 6 h. The purified biodiesel had 91.8 wt% of fatty acid ethyl esters.

- Hydroesterification is innovation biodiesel production using low-cost feedstocks by two steps: hydrolysis of waste cooking oil to FFAs in subcritical water and chemical esterification of FFAs into FAMEs (biodiesel) with methanol. In their paper, Santos and collaborators [21] performed an effective hydrolysis and validated the relationship between FFA and FAME yields; the study with experimental factorial design using a batch reactor with different parameters to obtain an optimal condition for the hydrolysis was used, and they found a conversion of the FFAs into 98.5 wt% of FAMEs with high quality by esterification in biodiesel production, suggesting that the biodiesel produced by the subcritical/chemical hydroesterification process has similar qualities to those of the biodiesel obtained by traditional alkaline transesterification.

- Recently (2018), Poppe et al. [22] described the use of an ultrasound system for the enzymatic transesterification of oils using combi-lipases as a biocatalyst. Reactions were performed in a mixture design of three factors to obtain the ideal mixture of lipases

(CALB, TLL, and RML) according to the composition of fatty acids present in each oil, and the main reaction variables were optimized. After 18 h of reaction, ultrasound provided a biodiesel yield of about 90% when using soybean oil and 70% using the waste oil. They concluded that having a good composition of fatty acids in soybean oil and waste oil the ultrasound technology can increase the yields of the reaction of biodiesel synthesis reaction. Using combi-lipase catalysts and ultrasound, the waste oil can be a good source of biodiesel to decrease the process costs in the enzymatic production of biodiesel.

Moreover, there are in the literature some papers about production of biodiesel on a pilot scale and socioeconomic impacts or life cycle assessment (LCA), as can be seen below:

Schneider et al. [23] aimed to recognize the production condition of waste oils and fats in the urban area of small towns (Arroio do Tigre-State of Rio Grande do Sul-5511 inhabitants) in order to define actions that will reduce the environmental impact of this activity in public places and homes. They analyzed the frying oil from two restaurants and produced biodiesel on a pilot scale with the $CH_3ONa$ catalyst and obtained 96.6% of conversion, and the quality parameters agree with those of the National Petroleum Agency (ANP).

In another paper, Moecke et al. [24] described a biodiesel production plant installed at Pinheira Beach in the state of Santa Catarina (Brazil), and the transesterification production process was used to convert waste cooking oils into biodiesel using basic catalysis in the presence of methanol. A batch system with the capacity to produce 200 L of biodiesel per day was adopted, and a 94.38% ester content was reached. They also discussed the environmental, social, and economic impacts by analyzing the biodiesel production by life cycle assessment (LCA). The LCA shows some categories that most contribute to the emissions in the biodiesel production process. With 92.10% of contribution, the collection stage is the most emitter with a huge difference in the production stages, that they send 7.9% only. This plant of biodiesel production can provide some gains for the local community like the environmental education and cooking oil recycling opportunities.

Recently, Silva Filho and collaborators [25] produced two integrated papers where they first produced biodiesel from waste frying oil and performed a kinetic study of the transesterification reaction for each frying oil source. After that, in a diesel oil engine, they measured energy and greenhouse gas emission efficiency values. Besides that, people answered some questions for the evaluation of the environmental and economic sustainability of the city of São Paulo. Their measurements showed that the biodiesel kinetics were zero order relative to the frying oil concentration, leading to a 90.10% yield, and the biodiesel physicochemical properties agree with the ANP standards. For blends up to 30%, there was a reduction in greenhouse gas emissions by 33% compared to diesel oil. According to the statistical survey, São Paulo has a production potential of more than 8800 $m^3$/month, which could generate savings of US$ 5,000,000/month or US$ 7,000,000/month considering the sale of excess biodiesel, glycerin, and carbon credits. In this second paper, they related the variation of the concentration of waste frying oil (WFO) with the reaction time and temperature during the transesterification of WFOs collected in São Paulo. Then, the biodiesel samples were mixed with the S-10 diesel oil in order to obtain the B10, B20, B30, B40, B50, B75, and B100 blends, which were tested in a diesel engine, and their power, fuel consumption, and gas emissions (CO, $CO_2$, and $SO_2$) have been measured to verify their greenhouse effect and energy efficiency [26].

For the worldwide famous city of Rio de Janeiro, a LCA study was performed by Souza et al. [27]; they analyzed the potential environmental impacts and socioeconomic benefits of installing a pilot plant for biodiesel production from used cooking oil (UCO) collected from hotels, restaurants, and bars in the Copacabana district of Rio de Janeiro, Brazil. An analysis was made of the production of biodiesel by the alkaline transesterification of UCO with ethanol using data taken from a pilot plant with a capacity to produce 250 l/batch. The findings of their economic analysis showed that it would be feasible for the Rio de Janeiro city council to implement a pilot plant in the Copacabana district of the city. The results of the social performance analysis indicated that the project would have a positive impact on local jobs and income generation.

Another study related to ambient aspects was developed in 2017 by D'Agosto et al. [28], in which there is an examination about the $CO_2$ emissions from the combustion of a biodiesel-diesel blend in stationary internal combustion engines to generate electricity by an original approach. They analyzed emissions according to the feedstock used for biodiesel production—soybean oil, palm oil, waste frying oil—through the methyl and ethyl routes. Besides that, the Tukey test was used, showing that, in general, higher engine loads led to a decrease in $CO_2$ emissions in comparison with the standard B4 (4% biodiesel) blend mandated in Brazil in that year.

### 2.2. Animal fats

Most researches related to renewable fuel are mainly focused on obtaining biodiesel obtained from vegetable oils. However, the use of a large supply of vegetable oils as feedstocks may lead to food scarcity because of the edible oil sources, which are primarily meant for human consumption. Consequently, the use of animal fat waste as cheap sources of feedstock in biodiesel production has gained more and more interest, including by Brazilian researchers [29].

#### 2.2.1. *Beef tallow*

Brazil owns one of the largest cattles in the world, and this helps to explain the fact that, since 2008, the use of beef tallow for biodiesel production has become increasingly significant, accounting for about 20% of sales of this biofuel in the country. [30]. For this reason, a diversity of scientific studies are found in the literature, as described below:

- In a paper of Araújo et al. [31], published in 2010, the transesterification of tallow by homogeneous catalysis with a solution of methanolic KOH was carried out by means of heating and preliminary formation of a microemulsion, and the beef tallow biodiesel was obtained with a high yield (96.26%). The fatty acids from beef tallow were characterized by gas chromatography-mass spectrometry (GC–MS) and quantified by thermogravimetric analysis (TGA). The correlation of data between the TGA and GC-MS techniques showed a good linear correlation coefficient, making it possible and appropriate for employment for compositional studies (%) of saturated and unsaturated fatty acids in samples of biodiesel from beef tallow by thermogravimetric techniques.
- In 2009, Teixeira and co-workers [32] reported that despite the use of ultrasonic energy for biodiesel production from different vegetable oils, its application for biodiesel production

from beef tallow has received little attention. Then, they performed the transesterification of beef tallow with methanol using ultrasound irradiation, and their results indicated that the reaction conversion and biodiesel quality were similar; however, the use of ultrasonic irradiation decreased the reaction time, showing that this method could be a promising alternative to the conventional method.

Considering another energy source, in 2012, Da Rós et al. [33] developed by a full $2^2$ factorial design leading to a set of seven runs with different combinations of molar ratio and temperature to find the optimal conditions for the microwave-assisted enzymatic (Burkholderia cepacia immobilized on silica-PVA) synthesis of biodiesel. Their main goal was to reduce the reaction time preliminarily established by a process of conventional heating. Under optimized conditions, almost total fatty acid present in the original beef tallow was converted into ethyl esters, representing an increase in sixfold for the process carried out under conventional heating.

- The nature of the solid residue formed in beef tallow biodiesel from two commercial producers in Brazil was determined by comparative analytical techniques, namely, gas chromatography with flame ionization detector (GC-FID) and thermogravimetry (TG) in a study of Fernandes Júnior and collaborators [34]. Their chromatographic and thermogravimetric results confirmed the nature of the residue as saturated monoglycerides, predominantly monostearin and monopalmitin.

- In another study, Moraes et al. [35] showed that the technique CGxCG can show results of characterization of oils and blends of the biodiesel/petrodiesel providing the percentage of the esters in biodiesel and the diesel blends.

- Even the biodegradation by microbial action was already studied. In their paper, Cazarolli and collaborators [36] compared the microbial growth (spores of the filamentous fungi *Pseudallescheria boydii*) in biodiesel obtained from tallow catalyzed by NaOH or KOH and purified with water or the solid adsorbent magnesium silicate. After 60 days, they related lower biomass formation in the biodiesel NaOH-water followed by biodiesel NaOH-magnesol and that the biodiesel KOH-magnesol and KOH-water favored the biomass formation.

Analogous to the waste oils, the beef tallow has also been studied in terms of its production in pilot scale, blends with diesel, and tests in diesel engines. For instance, Cunha and co-workers [37] produced biodiesel from beef tallow in a pilot plant with the capacity of approximately 800 kg/day. The main conclusion of this work was that the alkaline transesterification of beef tallow with methanol produces a biodiesel with high quality and also with a good conversion ratio, as well as the process is possible but the economical viability must be improved by recovering methanol and glycerol. The obtained results have been used by them for industrial scale-up of the process. Previously, some members of the same research group have been the authors of a study reporting that blends of diesel/biodiesel were produced and assessed in relation to some combustible properties. Their produced blends and biodiesel (B100) were also compared with diesel through consumption tests in a diesel engine used for energy generation, and all tests demonstrated that biodiesel and its formulations with diesel can present similar or better results to those of mineral diesel [38].

In 2011, Corrêa et al. [39] investigated the effect of the blend of biodiesel from beef tallow (B5) in commercial diesel oil on engine performance, analyzing possible internal consequences and characteristics of lubricating oil after the prolonged use in a diesel engine. The engine was operated for 600 hours and its performance was evaluated through tractor power take off (PTO) tests. After the end of the test, the engine performance was satisfactory although there was a tendency of reduction in power and increase in fuel consumption along the 600 hours. In another study with engine tests, Pereira and collaborators [40] realized several tests using beef tallow and beef tallow biodiesel in energy generation. There is also discussion about environmental, social, and economic aspects. Their results for performance and emissions of using beef tallow and beef tallow biodiesel in a stationary engine show, among others, that beef tallow can be successfully applied in compression ignition engine blending with diesel up to 15% of beef tallow. Moreover, the produced biodiesel can be applied in compression ignition engines neatly or blended with diesel in any proportion, with reservations to be used in cold weather conditions when in neat form, due to its relatively low pour point.

In order to obtain a biodiesel with the quality parameters required by the National Petroleum Agency (ANP), several researchers have studied the application and properties of biodiesel blends of beef tallow with biodiesel from other sources. There are also some articles that make a comparative study between biodiesel of animal and vegetable origin.

For instance, Canesin et al. [41] reported that when the parameters of quality of residual bovine and chicken oils were tested, some better points than those of the vegetable oil traditional for the biodiesel production were obtained. While they show a higher concentration of monosaturated and saturated fatty acids with a similar reduction in polyunsaturated fatty acids from soybean, the point of combustion, the flash point, and the density were higher than those of the petroleum diesel. Other parameters were measured like the iodine value that indicates their stability during the storage and showed that the biodiesel from residual oil had satisfactory values that can be improved during the production.

In another study, Teixeira et al. [30] prepared and monitored the quality of blends of beef tallow biodiesel with soybean biodiesel and with conventional mineral diesel fuel to study ideal proportions of the fuels. Measuring the viscosity, density, cold filter plugging point, and flash point (**Table 1**), they demonstrated that tallow biodiesel can be blended with both mineral diesel and soybean biodiesel to improve the characteristics of the blend fuels, over those of the tallow.

| Sample | Viscosity at 40°C, $mm^2\ s^{-1}$ | Density at 20°C, $kg\ m^{-3}$ | Cold filter plugging point, °C | Flash point, °C |
|---|---|---|---|---|
| Beef tallow biodiesel | 4.89 | 832.0 | 15 | 152 |
| Soybean biodiesel | 4.20 | 845.0 | 4 | 156 |
| Petrodiesel | 3.47 | 801.4 | 10 | 57 |

**Table 1.** Results of physicochemical tests performed on beef tallow biodiesel, soybean biodiesel, and petrodiesel samples [32].

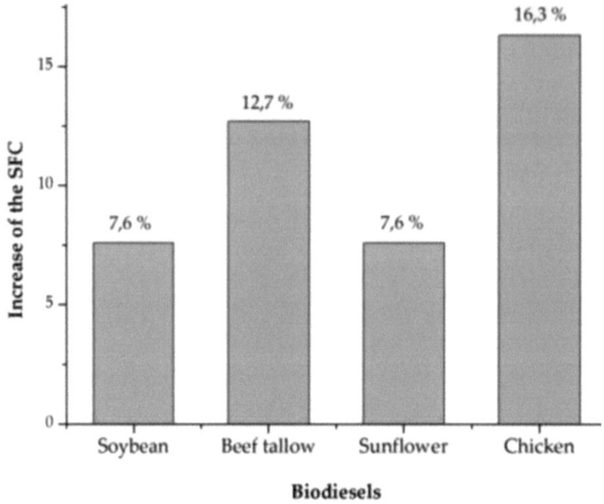


**Figure 2.** Comparison of the SFC using biodiesels related to the diesel oil [44].

On the other hand, Silva and co-workers [42] evaluated the engine generator (with a power of 7.36 kW on an electric generator with a power of 5.5 kW) performance using biodiesels obtained from soybean, sunflower, chicken fat, and beef tallow and compared them to mineral diesel. When (**Figure 2**) the specific fuel consumption (SFC) and the efficiency (Z) between the diesel oil with the biodiesel from sunflower and soybean oils were compared, biofuels showed a higher performance than the diesel, showing a lower increase of SFC (7.6%) and a slight increase in the overall efficiency.

An interesting approach was made by Pereira et al. [43] who studied an alternative to improve the oxidative quality of vegetable oil biodiesel by blending it with animal fat biodiesel. They studied the oxidative degradation of soybean/beef-tallow biodiesel (SB) 70/30 and 50/50 (w/w) during long-term storage (up to 350 days). The samples were stored and analyzed periodically via oxidative stability, among other analyses and the results showed that the produced blends presented higher oxidative stabilities than soybean biodiesel. During their long-term storage, the biodiesel blends showed greater resistance to oxidative degradation, which was indicated by the lower formation of hydroperoxides and polar compounds. Similarly, the decline in the tocopherol content was slower in SB biodiesel. Then, the prepared blends were appropriate alternatives to improve the oxidative quality of this biofuel. Referring to beef tallow and soybean biodiesel, Soldi et al. [44] synthesized sulfonated polystyrene (SPS) compounds from linear polystyrene (PS), resulting in catalytically active polymeric materials with several sulfonation degrees of dry polymer. The transesterification of beef tallow and soybean oil with ethanol and methanol was used to evaluate their catalytic performance. The SPS samples were insoluble in the reaction media and led to conversions of 85 and 75% of refined soybean oil and beef tallow with a 53 mg KOH/g acid number, respectively. There was an increase in ester production for reactions carried out at a higher methanol:oil molar ratio, and in some cases, products with as much as 94% methyl ester content were obtained, showing that the prepared sulfonated polymeric compounds are efficient catalysts for the these reactions.

Another comparative study was performed by Sales et al. [45], in which ethylic transesterification without any chemical or biochemical catalysts at different subcritical thermodynamic conditions using wet animal fat, soybean oil, and palm oil as feedstocks was realized. Conversions of 100, 84, and 98.5% were obtained for animal fat, soybean oil, and palm oil, respectively. Their results indicate that the process is energetically favorable, representing a cleaner and more advanced technology with environmental advantages for these processes. In another study, Teixeira et al. [46] have determined the physicochemical properties of biodiesel samples produced from mixtures of beef tallow, babassu oil, and soybean oil. The thermo-oxidative stability was evaluated by thermogravimetry (TG/DTG) and differential scanning calorimetry (DSC), and the results showed that the biodiesel obtained from a mixture containing 50% of babassu oil had lower values of pour point, cold filter plugging point, and freezing point.

To finish this sub-section, another comparative paper describes the evaluation of the potential of nonedible feedstocks to yield biodiesel by an enzymatic route (microbial lipase from *Burkholderia cepacia* immobilized on a silica-polyvinyl alcohol matrix). The ethanolysis of native oils from tropical crops, such as andiroba (*Carapa guianensis*), babassu (*Orbignya* sp.), jatropha (*Jatropha curcas*), macaw palm (*Acrocomia aculeata*), palm tree (*Elaeis guineensis*), and beef tallow in a solvent-free system and the biocatalyst were efficient in converting all fatty acids into the corresponding ethyl esters [47].

### 2.2.2. *Other waste animal lipids*

Despite the importance of beef tallow, reports about the use of chicken or swine fat, as well as fish oils have also been found in the literature as follows:

- In a study of 2008, Gomes et al. [48] determined certain physical-chemical characteristics of chicken oil that can influence the transesterification reactions of chicken oil. They also evaluated the potential for the production of chicken oil in the cooperatives of the western Paraná State and concluded that with the obtained yield of 95%, it would be possible by these cooperatives to produce up to 74,000 kg of biodiesel.day$^{-1}$.

- Another study involving chicken was developed by Cunha Jr. et al. [49], in which they optimized the conversion of a mix of chicken and swine fat residues for biodiesel production by a full $3^3$ factorial design for process parameters and analyzed their data using response surface methodology. After that, the optimum conditions were applied in a bench-scale reactor and the biofuel produced was characterized (except for oxidative stability and total glycerin, it agreed with quality requirements established by Official Regulations (ASTM 6751 and EN 14214), and it was observed that at high temperatures (50 and 70°C), phase separation between biodiesel and glycerol was impaired. Although high conversion was achieved (96.2%) at 70°C, this condition is not recommended because no spontaneous phase separation was verified. On the other hand, 30°C was the best temperature for biodiesel ethanolysis, with which they achieved around 83% conversion.

An alternative interesting animal source that has been studied by Brazilian researchers to produce biodiesel is the fish oil, once this country possessed large watersheds and an extensive coast. So, following we have some examples of these studies.

Almeida and Spanish collaborators [50] studied the influence of waste fish oil, palm oil, and waste frying oil as raw materials on biodiesel properties. They obtained biodiesel samples with yield higher than 82%, reaching 90% for palm oil (33.3 wt.%) and waste frying oil (66.7 wt.%) biodiesel. The FAME content was higher than 92.3% and had a maximum of 98.5% for waste fish oil (33.3 wt.%) and palm oil (66.7 wt.%) biodiesel. Their multi-objective optimization evidenced that although the use of the pure oils as feedstocks presented more advantages to biodiesel properties, the waste fish oil (42.1 wt.%) and waste frying oil (57.9 wt.%) mix is beneficial.

On the other hand, Mothé et al. [51] determined the optimal reaction conditions for transesterification of waste frying oil and fish by varying the reaction time, amount of catalyst, and temperature and have determined which of these variables exert a greater influence on the reaction yield. So, they found that the amount of catalyst was this variable for the transesterification of waste of the fish oil. So, the produced biodiesel was characterized using thermal analysis techniques and FTIR. The thermogravimetric technique was shown to be an important tool for calculating the yield of transesterification reactions, showing that the majority of TG reactions showed a yield exceeding 90%, a promising result while the FTIR spectra of biofuels obtained from frying and fish oils showed the characteristic regions of esters.

Among several fish species available in the market, the Nile Tilapia (*Oreochromis niloticus*) has an important role as a model of study probably due its rapid growth in fish farms and resistance to poor water quality and diseases, as can be seen in some studies, such as that realized by Santos et al. [52]. In that study, they evaluated the production of methyl esters from (Nile tilapia) oil and methanol using low-frequency high-intensity ultrasound (40 kHz) and response surface methodology (RSM) Their results showed that the most important operating condition affecting the reaction was the alcohol to FFA molar ratio, and it was possible to get a maximum yield of 98.2% after 90 min of reaction.

Another study involving Tilapias oil was performed by Martins and collaborators [53], which assessed mandatory parameters regulated by the Brazilian National Agency of Petroleum, Natural Gas and Biofuels (ANP) for biodiesel from residual tilapia oil. Then, they determined the values of specific mass, kinematic viscosity, water content, acidity level, flash point, and oxidative stability and the calorific value of fish oil biodiesel. Analysis of the biodiesel calorific value reported levels that are similar to those of diesel and indicated that fish oil is a promising alternative for biodiesel manufacturing. The same research group related in another work the evaluation of the yield and acid number of tilapia oil according to the type of waste used as well as the estimation of its potential for biodiesel production as a function of the oil obtained. They concluded that fish oil has potential (up to 370 L of biodiesel per ton of waste processed) to convert waste into biodiesel [54].

About the fish viscera oil, Rodrigues and co-workers [55] reported a study that describes the determination of kinetic and thermodynamic parameters of the degradation process of the biodiesel from this raw material. They performed several analyses of accelerated oxidation of biodiesel (EN 14112) and considered the reaction to be of the first order. Their results showed that the biodiesel oxidation reaction is non-spontaneous ($\Delta G^{\neq} > 0$) and endothermic ($\Delta H^{\neq} > 0$), and that the temperature and the oxidation concentration influence significantly the degradation process of methyl esters.

### 2.3. Different waste raw materials or applications of waste valorization for biodiesel production

There are other interesting wastes that have been studied for biodiesel production; among them, it is possible to highlight the soap stocks and waste refining oils, as can be seen in some papers, as follows. Soares et al. [56] investigated a new hydroesterification strategy for the production of biodiesel from soybean soap stock acid oil: complete hydrolysis in subcritical water, followed by the use of a packed-bed reactor, containing a fermented solid (Burkholderia cepacia LTEB11) with lipase activity, to obtain and convert the fatty acids to their ethyl esters. When the packed-bed reactor was reused in successive 48-h esterification reactions, they obtained conversions of over 84% of the fatty acids to esters.

Reis and co-workers [57] used three different high-acid-content soapstocks (soybean, palm, and coconut) to be esterified with short-chain alcohols in the presence of Amberlyst-35 (A35) sulfonic resin as a heterogeneous acid catalyst and p-toluenesulfonic acid as a homogeneous catalyst for comparison. Then, they noted that an increase in the hydrocarbon chain decreased the rate of conversion and the microwave irradiation reduced the reaction time from 6 to 1 h and achieved a maximum biodiesel production from coconut and soybean soap stocks and methanol (96–98%). Aguiar and collaborators [58] also used A35 to evaluate and to be compared to Amberlyst-36 and a new sulfonic resin, poly(divinylbenzene), synthesized by them in the laboratory (A-36) as to the catalytic activity for esterification of residue of the palm oil and soybean oil refining industry to produce biodiesel. In that paper, the resin A-36 exhibited catalytic activity similar to that of commercial resin A-35; the best results (87 and 94% for oleic and palmitic acid conversion) were achieved with the new sulfonic resin. When applied for the residue of the palm and soybean oil refining industry, all catalysts showed good conversions (76 and 93%) for all catalysts studied. This was the first example of using DVB sulfonic resins as catalysts in the esterification of fatty acid residues of the palm and soybean oil refining industry.

When palm oil is refined, it generates a residue that corresponds to 4% of the product process called the distillate of deodorization of palm oil (DDOP). The composition of the waste produced in free fatty acids (83%) ensures that they are a good feedstock for biodiesel synthesis [59]. In a study of Nascimento et al. [11], they used a catalyst prepared from a waste material (Amazon flint kaolin) for the esterification of DDPO with methanol, obtaining a maximum esterification activity of 92.8%, affirming that DDPO using that catalyst could be a cheaper alternative for production of biofuels. In sequence, the same research group reported a study where Pires et al. [60] incorporated 12-tungstophosphoric acid (HPW) into kaolin waste (MK700), MCM- 41, MCM-48, and SBA-15, and these obtained solid acid catalysts were used in the esterification reaction of DDPO with ethanol. Their results indicated that HPW supported on kaolin waste (as well as in MCM-41, MCM-48, and SBA-15) that was reaching 83% conversion at 2 h reaction with a 1:10 (DDPO:ethanol) molar ratio can provide significant advances in the development of environmentally benign processes in the biodiesel production.

Another residue was studied by Oliveira et al. [61]. They reported the esterification using waste oil generated in the refining of coconut oil with ethanol and methanol, performing it with and without adsorption of water in order to verify the effect of removing water on the reaction conversion, and conversions of over 99% were observed. Some reactions of synthesis of biodiesel require a huge quantity of álcool for a high yield. An experiment performed with an excess of 200% of álcool in a reaction using a zeolite 3A like adsorbent showed higher yield than the reaction with 800% of excess álcool in a reaction without water adsorption, proving to be a good candidate to be used in the biodiesel industry.

About the use of wastes for biodiesel production, there are still some different reports in the literature; for example, there is one paper that investigated the potential that the yeast *Candida lipolytica* UCP0988, in an anamorphous state, has to produce simultaneously a bioemulsifier and to accumulate lipids using inexpensive and alternative substrates such as waste soybean oil, and one of the biomass obtained was able to accumulate lipids of 0.425 g/g biomass (corresponding to 42.5%), which consisted of palmitic acid (28.4%), stearic acid (7.7%), oleic acid (42.8%), linoleic acid (19.0%), and linolenic acid (2.1%) [62]. In another interesting study, Jesus et al. [63] showed the use of vegetable waste as a potential source of raw material for biofuel production. In the work, the microalga *Chlorella vulgaris* was used to maximize the production of lipids and carbohydrates, and the culture medium was prepared with vegetable waste (pumpkin, zucchini, potato, eggplant, broccoli, carrot, cabbage, tomato, and green bean.) The positive results of this work motivate us to replace traditional culture media with media prepared with vegetable waste.

At last, Gomes & Pasquini [64] described the purification of sunflower-oil-derived biodiesel by dry cold washing using chicken's eggshells (*Gallus Gallus domesticus*), as well as a subsequent comparison with conventional wet washing with hot water and the reuse of eggshells. Both purification methods were efficient in the removal of free glycerin, free fatty acid molecules, and the catalyst. Thus, they provide the advantage of reducing effluent emission as well as offering an application for eggshells of which large amounts are generated due to excessive egg consumption and for which an appropriate use has not yet been found.

## 3. Conclusions

In this chapter, it was possible to understand the importance of waste valorization for the future of biodiesel production, either by using waste cooking oils and beef tallow or by investigating new techniques for analysis and catalysts and methods of synthesis or by the development of pilot plants and economic studies. If these materials cannot replace totally edible oils or diesel, it is clear that their use could be very higher in the energy matrix. For this reason, all researches have great importance, since those that could help to improve the social inclusion and minimize poverty, even those in which there is an improvement of the biodiesel production, by their properties, yields, process, as well as the production of catalysts more efficient, environmental friendly, among others technical aspects. Considering the highlighted position of Brazil in the world scenario of biofuels (despite the scarcity of research funds in this country), it is also possible to affirm that these researches gave/give a fundamental contribution to this reality, improving methodologies, investigating new types of wastes, upgrading their chain of collect and valorization by sensitizing and including more and more people. At the end, it is possible to conclude that the researched ones developed in Brazil occupy a position of vanguard in the world scene of the researches that involve the use of wastes for the production of biofuels.

## Conflict of interest

The authors declare no conflict of interest.

## Author details

Luís Adriano Santos do Nascimento*, Deborah Terra de Oliveira, Alex Nazaré de Oliveira, Luiza Helena de Oliveira Pires, Carlos Emmerson Ferreira da Costa and Geraldo Narciso da Rocha Filho

*Address all correspondence to: adrlui1@yahoo.com.br

Federal University of Pará, Belém, Brazil

## References

[1] IEA World Energy Outlook 2008, 2009 and 2010

[2] Styring S. Artificial photosynthesis for solar fuels. Faraday Discussions. 2012;**155**:357-376. DOI: 10.1039/C1FD00113B

[3] Lewis NS, Nocera DG. Powering the planet: Chemical challenges in solar energy utilization. Proceedings of the National Academy of Sciences of the United States of America. 2006;**103**:15729-15735. DOI: 10.1073/pnas.0603395103

[4] ANP. Anuário Estatístico Brasileiro Do Petróleo, Gás Natural E Biocombustíveis. Agência Nac Do Petróleo, Gás Nat e Biocombustíveis 2016. http://www.anp.gov.br/images/publicacoes/Anuario_Estatistico_ANP_2016.pdf

[5] Rogelj J et al. Paris Agreement climate proposals need a boost to keep warming well below 2°C. Nature. 2016;**534**:631-639. DOI: 10.1038/nature18307

[6] Festel G et al. Modelling production cost scenarios for biofuels and fossil fuels in Europe. Journal of Cleaner Production. 2014;**66**:242-253

[7] Araújo CDM, Andrade CC, Silva ES, Dupas FA. Biodiesel production from used cooking oil: A review. Renewable and Sustainable Energy Reviews. 2013;**27**:445-452

[8] Cruz MG, Guerreiro E, Raiher AP. A Evolução da Produção de Etanol no Brasil, no Período de 1975 a 2009. Documentos técnicos-científicos. 2012;**43**:142-159

[9] De Oliveira FC, Coelho ST. History, evolution, and environmental impact of biodiesel in Brazil: A review. Renewable and Sustainable Energy Reviews. 2017;**75**:168-179. DOI: 10.1016/j.rser.2016.10.060

[10] de RC L, Leal MR. O biocombustível no Brasil. Novos estudos CEBRAP. 2007;**78**:15-21. DOI: 10.1590/S0101-33002007000200003

[11] Nascimento LAS, Angélica RS, Costa CEF, Zamian JR, Rocha Filho GN. Conversion of waste produced by the deodorization of palm oil as feedstock for the production of biodiesel using a catalyst prepared from waste material. Bioresource Technology. 2011; **102**:8314-8317. DOI: 10.1016/j.biortech.2011.06.004

[12] Loures CCA, Amaral MS, Da Rós PCM, Zorn SMFE, De Castro HF, Silva MB. Simultaneous esterification and transesterification of microbial oil from Chlorella minutissima by acid catalysis route: A comparison between homogeneous and heterogeneous catalysts. Fuel. 2018;**211**:261-268. DOI: 10.1016/j.fuel.2017.09.073

[13] Aboim JB, Oliveira DT, Ferreira JE, Siqueira AS, Dall'agnol LT, Rocha Filho GN, et al. Determination of biodiesel properties based on a fatty acid profile of eight Amazon cyanobacterial strains grown in two different culture media. RSC Advances. 2016;**6**: 109751-109758. DOI: 10.1039/C6RA23268J

[14] Lima RP, Da Luz PTS, Braga M, Batista PRS, Costa CEF, Zamian JR, et al. Murumuru butter and oils of buriti and pracaxi can be used for biodiesel production: Physico-chemical properties and thermal and kinetic studies. Industrial Crops and Products. 2017;**97**:536-544. DOI: 10.1016/j.indcrop.2016.12.052

[15] César AS, Werderits DE, Saraiva GLO, Guabiroba RCS. The potential of waste cooking oil as supply for the Brazilian biodiesel chain. Renewable and Sustainable Energy Reviews. 2017;**72**:246-253. DOI: 10.1016/j.rser.2016.11.240

[16] Costa Neto PR, Rossi LFS, Zagonel GF, Ramos LP. Produção de biocombustível alternativo ao óleo diesel através da transesterificação de óleo de soja usado em frituras. Quim Nova. 2000;**23**:531-537

[17] Hocevar L, Soares VRB, Oliveira FS, Korn MGA, Teixeira LSG. Application of Multivariate Analysis in Mid-Infrared Spectroscopy as a Tool for the Evaluation of Waste Frying Oil Blends. Journal of the American Oil Chemists' Society. 2012;**89**:781-786. DOI: 10.1007/s11746-011-1968-8

[18] Da Silva ML, Figueiredo AP, Cardoso AL, Natalino R, Da Silva MJ. Effect of water on the ethanolysis of waste cooking soybean oil using a tin(II) chloride catalyst. Journal of the American Oil Chemists' Society (JAOCS). 2011;**88**:1431-1437. DOI: 10.1007/s11746-011-1794-z

[19] Alberici RM, Souza V, Sá GF, Morelli SR, Eberlin MN, Daroda RJ. Used frying oil: A proper feedstock for biodiesel production? Bioenergy Research. 2012;**5**:1002-1008. DOI: 10.1007/s12155-012-9216-0

[20] Vescovi V, Rojas MJ, Baraldo A, Botta DC, Santana FAM, Costa JP, et al. Lipase-catalyzed production of biodiesel by hydrolysis of waste cooking oil followed by esterification of free fatty acids. Journal of the American Oil Chemists' Society (JAOCS). 2016;**93**:1615-1624. DOI: 10.1007/s11746-016-2901-y

[21] Santos LK, Hatanaka RR, Oliveira JE, Flumignan DL. Experimental factorial design on hydroesterification of waste cooking oil by subcritical conditions for biodiesel production. Renewable Energy - Journal. 2017;**114**:574-580. DOI: 10.1016/j.renene.2017.07.066

[22] Poppe JK, Matte CR, Fernandez-Lafuente R, Rodrigues RC, Ayub MAZ. Transesterification of waste frying oil and soybean oil by combi-lipases under ultrasound-assisted reactions. Applied Biochemistry and Biotechnology. 2018;**18**:1-14. DOI: 10.1007/s12010-018-2763-x

[23] Schneider RCS, Santos E, Krise DJ, Lipke RJ. Residual fatty and oil production in Arroio do Tigre Town, Rio Grande do Sul, aiming production. Acta Scientiarum Technology. 2011;**33**:71-79. DOI: 10.4025/actascitechnol.v33i.1.8823

[24] Moecke EHS, Feller R, Santos HAD, Machado MDM, Cubas ALV, Dutra ARDA, et al. Biodiesel production from waste cooking oil for use as fuel in artisanal fishing boats: Integrating environmental, economic and social aspects. Journal of Cleaner Production. 2016;**135**:679-688. DOI: 10.1016/j.jclepro.2016.05.167

[25] Silva Filho SCD, Miranda AC, Silva TAF, Calarge FA, Souza RRD, Santana JCC, et al. Data on kinetic, energy and emission performance of biodiesel from waste frying oil. Data Br. 2018;**18**:1224-1228. DOI: 10.1016/j.dib.2018.04.017

[26] Silva Filho SCD, Miranda AC, Silva TAF, Calarge FA, Souza RRD, Santana JCC, et al. Environmental and techno-economic considerations on biodiesel production from waste frying oil in São Paulo city. Journal of Cleaner Production. 2018;**183**:1034-1042. DOI: 10.1016/j.jclepro.2018.02.199

[27] Souza DDP, Mendonça FM, Nunes KRA, Valle R. Environmental and socioeconomic analysis of producing biodiesel from used cooking oil in Rio de Janeiro: The case of the Copacabana district. Journal of Industrial Ecology. 2012;**16**:655-664. DOI: 10.1111/j.1530-9290.2012.00517.x

[28] D'Agosto MA, Da Silva MAV, Franca LS, Oliveira CM, Alexandre MOL, Marques LGC, et al. Comparative study of emissions from stationary engines using biodiesel made from soybean oil, palm oil and waste frying oil. Renewable and Sustainable Energy Reviews. 2017;**70**:1376-1392. DOI: 10.1016/j.rser.2016.12.040

[29] Alajmi FSMDA, Hairuddin AA, Adam NM, Abdullah LC. Recent trends in biodiesel production from commonly used animal fats. International Journal of Energy Research. 2017:885-902. DOI: 10.1002/er.3808

[30] Teixeira LSG, Couto MB, Souza GS, Filho MA, Assis JCR, Guimarães PRB, et al. Characterization of beef tallow biodiesel and their mixtures with soybean biodiesel and mineral diesel fuel. Biomass and Bioenergy. 2010;**34**:438-441. DOI: 10.1016/j.biombioe.2009.12.007

[31] Araújo BQ, Nunes RCR, Moura CVR, Moura EM, Citó AMGL, Dos Santos Júnior JR. Synthesis and characterization of beef tallow biodiesel. Energy and Fuels. 2010;**24**:4476-4480. DOI: 10.1021/ef1004013

[32] Teixeira LSG, Assis JCR, Mendonça DR, Santos ITV, Guimarães PRB, Pontes LAM, et al. Comparison between conventional and ultrasonic preparation of beef tallow biodiesel. Fuel Processing Technology. 2009;**90**:1164-1166. DOI: 10.1016/j.fuproc.2009.05.008

[33] Da Rós PCM, De Castro HF, Carvalho AKF, Soares CMF, De Moraes FF, Zanin GM. Microwave-assisted enzymatic synthesis of beef tallow biodiesel. Journal of Industrial Microbiology & Biotechnology. 2012;**39**:529-536. DOI: 10.1007/s10295-011-1059-8

[34] Fernandes Júnior VJ, Araujo ADS, Vinhado FDS, Pivesso PR. Caracterização de resíduo sólido formado em biodiesel de sebo bovino. Quim Nova. 2012;**35**:1901-1906. DOI: 10.1590/S0100-40422012001000002

[35] Moraes MSA, Zini CA, Gomes CB, Bortoluzzi JH, Von Mühlen C, Caramão EB. Uso da cromatografia gasosa bidimensional abrangente (gc×gc) na caracterização de misturas biodiesel/diesel: Aplicação ao biodiesel de sebo bovino. Quim Nova. 2011;**34**:1188-1192. DOI: 10.1590/S0100-40422011000700016

[36] Cazarolli CJ, Bücker F, Manique MC, Krause LC, Maciel GPS, Onorevoli B, et al. Susceptibility of biodiesel from tallow to biodegradation by Pseudallescheria boydii. Revista Brasileira de Biociências. 2012;**10**

[37] Cunha da ME, Krause LC, Moraes MSA, Faccini CS, Jacques RA, Almeida SR, et al. Beef tallow biodiesel produced in a pilot scale. Fuel Processing Technology. 2009;**90**:570-575. DOI: 10.1016/j.fuproc.2009.01.001

[38] Moraes SMA, Krause LC, Cunha ME, Faccini CS, Menezes EW, Veses RC, et al. Tallow biodiesel: Properties evaluation and consumption tests in a diesel engine tallow biodiesel: Properties evaluation and consumption tests in a diesel engine. Energy & Fuels. 2008;**22**: 1949-1954. DOI: 10.1021/ef7006535

[39] Corrêa IM, Maziero JVG, Storino M. Mistura de biodiesel de sebo bovino em motor diesel durante 600 horas. Ciência Rural. 2011;**41**:1189-1194. DOI: 10.1590/S0103-84782011005000088

[40] Pereira RG, Tulcan OEP, Fellows CE, Lameira VJ, Quelhas OLG, Aguiar ME, et al. Sustainability and mitigation of greenhouse gases using ethyl beef tallow biodiesel in energy generation. Journal of Cleaner Production. 2012;**29–30**:269-276. DOI: 10.1016/j.jclepro. 2012.01.007

[41] Canesin EA, de Oliveira CC, Matsushita M, Dias LF, Pedrão MR, de Souza NE. Characterization of residual oils for biodiesel production. Electronic Journal of Biotechnology. 2014; **17**:39-45. DOI: 10.1016/j.ejbt.2013.12.007

[42] Da Silva MJ, Souza SNM, Chaves LI, Rosa HA, Secco D, Santos RF, et al. Comparative analysis of engine generator performance using diesel oil and biodiesels available in

Paraná State, Brazil. Renewable and Sustainable Energy Reviews. 2013;**17**:278-282. DOI: 10.1016/j.rser.2012.09.037

[43] Pereira GG, Garcia RKA, Ferreira LL, Barrera-Arellano D. Soybean and soybean/beef-tallow biodiesel: A comparative study on oxidative degradation during long-term storage. Journal of the American Oil Chemists' Society (JAOCS). 2017;**94**:587-593. DOI: 10.1007/s11746-017-2962-6

[44] Soldi RA, Oliveira ARS, Ramos LP, César-Oliveira MAF. Soybean oil and beef tallow alcoholysis by acid heterogeneous catalysis. Applied Catalysis A: General. 2009;**361**:42-48. DOI: 10.1016/j.apcata.2009.03.030

[45] Sales EA, Ghirardi ML, Jorquera O. Subcritical ethylic biodiesel production from wet animal fat and vegetable oils: A net energy ratio analysis. Energy Conversion and Management. 2017;**141**:216-223. DOI: 10.1016/j.enconman.2016.08.015

[46] Teixeira GAA, Maia AS, Rosenhaim R, Santos IMG, Souza AL, Souza AG, et al. Thermo-oxidative decomposition of biodiesel samples obtained from mixtures of beef tallow, soybean oil, and babassu oil. Journal of Thermal Analysis and Calorimetry. 2011;**106**:569-574. DOI: 10.1007/s10973-011-1372-5

[47] Carvalho AKF, Da Rós PCM, Teixeira LF, Andrade GSS, Zanin GM, Castro HF. Assessing the potential of non-edible oils and residual fat to be used as a feedstock source in the enzymatic ethanolysis reaction. Industrial Crops and Products. 2013;**50**:485-493. DOI: 10.1016/j.indcrop.2013.07.040

[48] Gomes LFS, Souza SNM, Bariccatti RA. Biodiesel manufactured with chicken oil. Acta Scientiarum Technology. 2008;**30**:57-62

[49] Cunha A Jr, Feddern V, De Prá MC, Higarashi MM, De Abreu PG, Coldebella A. Synthesis and characterization of ethylic biodiesel from animal fat wastes. Fuel. 2013;**105**:228-234. DOI: 10.1016/j.fuel.2012.06.020

[50] Almeida VF, García-Moreno PJ, Guadix A, Guadix EM. Biodiesel production from mixtures of waste fi sh oil, palm oil and waste frying oil: Optimization of fuel properties. Fuel Processing Technology. 2015;**133**:152-160. DOI: 10.1016/j.fuproc.2015.01.041

[51] Mothé CG, De Castro BCS, Mothé MG. Characterization by TG/DTG/DSC and FTIR of frying and fish oil residues to obtain biodiesel. Journal of Thermal Analysis and Calorimetry. 2011;**106**:811-817. DOI: 10.1007/s10973-011-1795-z

[52] Santos FFP, Malveira JQ, Cruz MGA, Fernandes FAN. Production of biodiesel by ultrasound assisted esterification of Oreochromis niloticus oil. Fuel. 2010;**89**:275-279. DOI: 10.1016/j.fuel.2009.05.030

[53] Martins GI, Secco D, Rosa HA, Bariccatti RA, Dolci BD, De Souza SNM, et al. Physical and chemical properties of fish oil biodiesel produced in Brazil. Renewable and Sustainable Energy Reviews. 2015;**42**:154-157. DOI: 10.1016/j.rser.2014.10.024

[54] Martins GI, Secco D, Tokura LK, Bariccatti RA, Dolci BD, Santos RF. Potential of tilapia oil and waste in biodiesel production. Renewable and Sustainable Energy Reviews. 2015;**42**: 234-239. DOI: 10.1016/j.rser.2014.10.020

[55] Rodrigues JS, Valle CP, Guerra PAGP, Rios MAS, Malveira JQ, Ricardo NMPS. Study of kinetics and thermodynamic parameters of the degradation process of biodiesel produced from fish viscera oil. Fuel Processing Technology. 2017;**161**:95-100. DOI: 10.1016/j.fuproc.2017.03.013

[56] Soares D, Pinto AF, Gonçalves AG, Mitchell DA, Krieger N. Biodiesel production from soybean soapstock acid oil by hydrolysis in subcritical water followed by lipase-catalyzed esterification using a fermented solid in a packed-bed reactor. Biochemical Engineering Journal. 2013;**81**:15-23. DOI: 10.1016/j.bej.2013.09.017

[57] Reis MC, Freitas FA, Lachter ER, Gil RASS, Nascimento RSV, Poubel RL, et al. Biodiesel production from fatty acids of refined vegetable oils by heterogeneous acid catalysis and microwave irradiation. Quimica Nova. 2015;**10**:1307-1312. DOI: 10.5935/0100-4042.20150163

[58] Aguiar VM, Souza LA, Galdino FS, Silva MMC, Teixeira VG, Lachter ER. Sulfonated poly (divinylbenzene) and poly (styrene-divinylbenzene) as catalysts for esteri fi cation of fatty acids. Renewable Energy. 2017;**114**:725-732

[59] Aranda DAG, Antunes OAC. Catalytic Process to the Esterification of Fatty Acids Present in the Acid Grounds of the Palm Using Acid Solid Catalysts; 2004. WO Patent 2004096962

[60] Pires LHO, de Oliveira AN, Monteiro OV Jr, Angélica RS, Costa CEF, Zamian JR, et al. Esterification of a waste produced from the palm oil industry over 12-tungstophosforic acid supported on kaolin waste and mesoporous materials. Applied Catalysis B: Environmental. 2014;**160–161**:122-128. DOI: 10.1016/j.apcatb.2014.04.039

[61] Oliveira JFG, Lucena IL, Saboya RMA, Rodrigues ML, Torres AEB, Fernandes FAN, et al. Biodiesel production from waste coconut oil by esterification with ethanol: The effect of water removal by adsorption. Renewable Energy. 2010;**35**:2581-2584. DOI: 10.1016/j.renene.2010.03.035

[62] Souza AF, Rodriguez DM, Ribeaux DR, Luna MAC, Lima e Silva TA, Silva Andrade RF, et al. Waste soybean oil and corn steep liquor as economic substrates for bioemulsifier and biodiesel production by Candida lipolytica UCP 0998. International Journal of Molecular Sciences. 2016;**17**:1-18. DOI: 10.3390/ijms17101608

[63] Jesus SS, Santana A, Ponce GHSF, Maciel FR. Potential use of vegetable waste for biofuel production. Journal of Chemical Technology and Biotechnology. 2017;**92**:90-99. DOI: 10.1002/jctb.5002

[64] Gomes MG, Pasquini D. Utilization of eggshell waste as an adsorbent for the dry purification of biodiesel. Environmental Progress & Sustainable Energy. 2018:1-7. DOI: 10.1002/ep.12870

Chapter 3

# Properties of Torrefied Palm Kernel Shell via Microwave Irradiation

Razi Ahmad, Mohd Azlan Mohd Ishak,
Nur Nasulhah Kasim and Khudzir Ismail

Additional information is available at the end of the chapter

http://dx.doi.org/10.5772/intechopen.81374

**Abstract**

This study describes the characteristic and thermal properties of torrefied palm kernel shell (PKS) by microwave irradiation pretreatment. The microwave power level (200, 300, 450, and 600 W) and processing time (4, 8, and 12 min) were used in this study. The pretreated samples were analyzed for mass and energy yield, calorific value, proximate and elemental composition, and thermal decomposition. Results showed that the characteristic of pretreated PKS was enhanced by increasing the microwave power level and processing the time. The oxygen content and O/C ratio of torrefied PKS were reduced by increasing the microwave power level. The carbon content of pretreated PKS, which was closed to the untreated MB coal properties with comparable calorific value, was obtained. The microwave power level of 450 W and processing time of 8 min were suitable to upgrade the PKS to a respectable quality feedstock. Thus, it can be concluded that the alteration in physical, chemical, and thermal properties of torrefied PKS discovered the potential of this feedstock to be applied in subsequent thermochemical conversion such as pyrolysis and gasification.

**Keywords:** pretreatment, torrefaction, microwave irradiation, palm kernel shell, gasification

## 1. Introduction

In recent years, the increasing emission of $CO_2$, $SO_x$, and $NO_x$ has become a concern on the utilization of the world energy [1]. In the midst of limited availability of fossil fuels and high level of air pollution, energy efficient technologies are gaining importance, and gasification, being a highly efficient technology, has received significant attention [2]. Currently, coal is the main feedstock in gasification and is expected to be applied as the energy resource for many decades ahead. However, this direction is difficult to achieve due to the increase in energy demand that had caused the shortage supply and reduction of coal [3]. Consequently, one of the approaches is to utilize the biomass in thermochemical conversion such as pyrolysis, liquefaction, and gasification. The traditional use of biomass has been restricted to cooking and heating purposes, which has affected adverse impacts such as land degradation and desertification. However, the current use of biomass with a high-quality energy carrier transformed from raw biomass for electricity and heat production can substantially reduce emissions from the conventional power plants. This ability to convert raw biomass into convenient energy carriers increases the interest on biomass use for energy purpose, especially the lignocellulosic biomass [4].

Biomass is one of the capable renewable energy sources and is applied as solid, liquid, and gas fuels [5]. The biomass is an appealing concern worldwide, because of its nonedible characteristic, carbon neutrality, and relative abundance. Moreover, the rising worries about the effects of $CO_2$ emissions from fossil fuels call for sustainable energy sources, such as biomass [6]. In Malaysia, oil palm residues are considered to be the most plentiful biomass and the greatest prospects for fuel generation. Malaysia produces about 47% of the world's palm oil source and can be reflected as one of the world's largest producers of palm oil. Therefore, Malaysia creates huge quantity of oil palm biomass including palm kernel shell (PKS), oil palm trunks, oil palm fronds, empty fruit bunches, and fibers as residues from harvesting and processing activities [7]. The PKS as one of the residues from oil palm industry generated about 4.19 MnT in 2016 [8]. Therefore, PKSs appear to have prominent capacities to become an alternative source of energy for the country.

However, the utilization of biomass, which is a renewable and environmental friendly resource during thermal conversion, imposed several problems. The untreated biomass has the drawbacks as follows:

- **i.** Low energy content
- **ii.** High moisture and oxygenated compound
- **iii.** Hygroscopic behavior
- **iv.** Poor grindability

Likewise, it is reasonably challenging for straight application of untreated PKS as raw material for fuel production such as gasification or pyrolysis. Typically, the palm plantations and their processing plants are located in rural areas. Thus, the untreated PKS is opened to fungal attack and biodegradation through storage and transportation. The high moisture content also can interrupt the thermal conversion process for energy production [9]. The low energy density of PKS, normally 18 MJ/kg, with high moisture content typically around 10 wt.% as a result of its hygroscopic character, is a weakness of biomass [10–12]. As shown by the previous researcher [13–15], highly oxygenated biomass with high O/C ratio will lower the gasification efficiency in contrast with low O/C feedstock such as coal. Consequently, these properties of the untreated PKS were associated with several problems in biomass thermal conversion such as in gasification. As a result, prior to gasification, it might be attractive to transform the biomass characteristics.

A pretreatment method prior to thermal conversion is required in direction to reduce some of the aforementioned problems. Thus, torrefaction pretreatment appears to be an effective route. The thermal pretreatment or torrefaction at low temperature between 200 and 300°C, which operated in the nonappearance of oxygen, upgraded the untreated feedstock to more value feedstock. Nitrogen is generally used as carrier gas to provide a nonoxidizing atmosphere in most laboratory tests. Since torrefaction is conducted at conditions similar to those of pyrolysis that usually takes place between 350 and 650°C, torrefaction has also been known as mild pyrolysis [16]. The pretreated biomass has high calorific value and carbon fraction with low moisture content and O/C ratio compared to the untreated or original material. The energy value of pretreated material will increase with increasing carbon fraction and calorific value [17].

The previous studies have also shown other advantages of this torrefaction pretreatment, such as it improves feedstock hydrophobicity, homogeneity, and grindability [4, 18]. Satpathy et al. [19] found that the torrefied wheat and barley straw are more hydrophobic and the moisture uptake is reduced by 61–68% under suitable torrefaction condition. Torrefaction of marula seeds and blue gum improves the higher heating value and energy content of the biomass. The weight loss also increases when the torrefaction temperature increases due to moisture removal and hemicellulose breakdown, which produce $H_2O$, CO, $CO_2$, and other hydrocarbons [20].

Torrefaction temperature is one of the important parameters in torrefaction pretreatment [4]. Ibrahim et al. [21] found that the lower temperature and shorter residence time were the best treatment to achieve good physical properties with a relatively high energy yield. When treated at these conditions, the softwood mixture had the highest energy (95%), followed by the hardwood mixture (80%), then willow (79%), and finally eucalyptus (75%). Increasing the severity of the torrefaction conditions greatly improved the physical characteristics of the torrefied biomass, in terms of grindability properties and hydrophobicity. The optimum temperatures were reliant on raw material, and consequently, the effects specify that careful optimization is necessary for all feedstock types to increase the advantages of torrefaction at the same time preserving an adequate energy yield. As pretreatment conditions became more severe between temperature of 250 and 300°C, this led to a more qualified and energy-dense solid fuel with higher fixed carbon content, increased calorific values, and reduced hydrogen and oxygen contents [22]. By increasing the torrefaction temperature, the weight loss increased and bulk density decreased. The torrefied wood samples improved solid fuel property with high fuel ratio, which are close to lignite coal [23]. Mamvura et al. [20] found that the nonoxidative conditions with low heating rates and shorter residence time resulted in the best torrefied biomass. The increase in HHV together with increase in energy density for torrefied marula seeds during investigation intended that it is potential to co-conversion with coal making it a promising biomass source. Therefore, the pretreated or torrefied biomass, which has been improved in energy density, hydrophobicity, and grindability, overcomes the weakness of untreated biomass and is then driven to be applied in thermochemical conversion [4, 16].

Most of the biomass torrefactions applied the conventional electric heater, while there is an alternative technology designated microwave irradiation. Microwave technology has expanded remarkable importance in the thermochemical pretreatment of waste materials, including biomass, waste cooking oil, scrap tires, and others. Innovative fields are being exposed in which microwave can be applied as an alternative source of heating. The application of microwave in waste treatment originated about two decades ago. Therefore, it can be considered at an early stage of enlargement [24]. Microwave irradiation is an electromagnetic irradiation in the range of wavelengths from 0.01 to 1 m and the equivalent frequency range of 0.3–300 GHz. Normally, the microwave reactors for chemical synthesis and all domestic microwave ovens operate at 2.45 GHz frequency, which corresponds to a wavelength of 12.25 cm. Microwave irradiation has attracted much attention in recent years due to the advantages associated with dielectric heating effects. Microwave dielectrics are known as a material, which absorbs microwave irradiation; thus, microwave heating is called dielectric heating [25]. The pretreatment using microwave irradiation is an effective method for upgrading the biomass [26]. Unlike conventional heating technique in which heat gradually enters into samples over normal heat transfer mechanisms (convection, conduction, and radiation) [27], microwave irradiation employs electromagnetic energy to produce heat, which can enter deep into samples, permitting heating to initiate volumetrically [28]. Microwave irradiation has many advantages such as:

- **i.** Noncontact heating
- **ii.** Energy transfer instead of heat transfer
- **iii.** Rapid heating
- **iv.** Selective material heating
- **v.** Volumetric heating
- **vi.** Quick start-up and stopping
- **vii.** Heating from the interior of the material body [25, 29]

Wang et al. [30] utilized microwave irradiation to upgrade the properties of rice husk and sugarcane residues by varying different parameters, including microwave power level and processing time. They found that the suitable microwave power levels are proposed to be set between 250 and 300 W for the torrefaction of these two agricultural wastes. Also, with appropriate processing time, the caloric value is able to increase 26% for rice husk and 57% for sugarcane residue. Huang et al. [31] found that higher microwave power levels contributed to higher heating rate and reaction temperature and therefore produced the torrefied biomass with higher heating value and lower H/C and O/C ratios. The torrefied biomass or biochar probably substitutes coal due to high heating value and fuel ratio as well as low atomic H/C and O/C ratios. The microwave torrefaction of *Leucaena* produced thermally stable biochar compared with sewage sludge at lower microwave power levels, which means that the microwave heating performance of *Leucaena* is better. Compared with conventional torrefaction, mass and energy yields of microwave torrefaction were lower, which might be attributable to the further severe reaction accomplished by microwave irradiation [32].

Consequently, more research is required to entirely understand the characteristic of torrefied biomass using microwave irradiation prior to further thermochemical conversion. It is also necessary to understand the thermal decomposition of torrefied biomass during pyrolysis in thermogravimetric analyzer (TGA) since, in the thermal conversion studies, the beginning stage involves the feedstock devolatilization. Therefore, in this study, the PKS was initially torrefied in microwave, and the properties of torrefied PKS were explored. Subsequently, the thermal decomposition and behavior of torrefied PKS during pyrolysis process using TGA were examined.

## 2. Method

### 2.1. Materials

PKS as a biomass sample was obtained from United Oil Palm Mill Sdn. Bhd., Nibong Tebal, Penang, Malaysia. The PKS is produced from the shell/kernel separator. The PKS sample was crushed and sieved through progressively finer screen to obtain particle sizes in the range of 200–400 μm. The untreated PKS sample was dried in an oven at 105°C for 24 h for rendering moisture-free and finally stored in an air-tight container until the experiments and analyses were carried out. The pre-drying is needed to avoid further biodegradation of the sample through storage since the moisture mass fraction of the raw PKS is relatively high [33]. Moreover, the pre-drying is used to simulate the industrial practice of sun-drying the materials before storage [6].

### 2.2. Torrefaction experiment

The torrefaction experiment was carried out in a domestic microwave oven (Samsung) with technical specifications of ~240 V and 50 Hz and a maximum power of 800 W. The microwave output power levels of 200, 300, 450, and 600 W were used in this study. The untreated PKS of 5 g was put in the sample crucible placed at the center of the microwave oven. Then, the nitrogen gas at a flow rate of 50 mL/min was purged in the reaction compartment to retain the inert atmosphere condition. After 10 min purging, the microwave system was turned on, and the microwave output power level was selected with respective processing time of 4, 8, and 12 min. The inert atmosphere condition was continued during the microwave irradiation. The power supply was turned off, and the nitrogen gas flow was stopped after the set processing time was achieved. The final temperature of the pretreated PKS was measured using infrared thermometer immediately after the pretreatment process. The final weight of pretreated PKS was measured once it reached the room temperature. The experiment under all of the studied parameters was repeated to confirm the measurement quality and repeatability of the achieved results.

### 2.3. Calculation of solid conversion, mass yield, and energy yield

The solid conversion ($X_s$), mass yield ($Y_m$), and energy yield ($Y_e$) of the pretreated samples were calculated according to Eqs. (1)–(3), respectively:

$$X_s = (M_u - M_p)/M_u \quad (1)$$

$$Y_m = (M_p/M_u) \times 100 \quad (2)$$

$$Y_e = Y_m \times (CV_p/CV_u) \quad (3)$$

where $M$ is the mass of sample, CV is the calorific value, the subscript $u$ means the value of untreated sample, and the subscript $p$ means the value of pretreated sample.

## 2.4. Sample analyses

The physical and chemical characteristics of the untreated and pretreated samples were analyzed. The elemental composition of the sample was examined using elemental analyzer CHNS-O Flash 2000. The elemental composition examines the carbon (C), hydrogen (H), nitrogen (N), and sulfur (S) contents. The oxygen content was analyzed by the different of total mass content. The sample of 2 mg was weighted and encapsulated into a tin capsule. The sample was placed in the sample loading chamber. During the analysis, the sample was dropped into a furnace held at 1000°C. At the same time the sample drops into the furnace, a dose of oxygen is released into the furnace. The sample was combusted by the heated oxygen-rich environment. The products of elemental analysis are $CO_2$, $H_2O$, $NO_x$, and $SO_x$. These gases, which were carried through the system by the helium carrier, will be swept through the oxidation tube packed with copper sticks (which removes oxygen), to complete the conversion to $SO_2$. These gases are passed through four infrared detectors of C, H, N, and S, and the results were displayed as weight percent of C, H, N, and S.

The proximate analysis that analyzed the moisture, volatile matter, ash, and fixed carbon content was carried out using a Mettler Toledo thermogravimetric analyzer (TGA) according to the standards of the American Society for Testing and Materials (ASTM). For each analysis, about 10 mg of sample was weighted using a microbalance and placed in a ceramic crucible. Next, this crucible was positioned in the furnace where the analysis was performed. The programmed TGA began by applying the heating rate of 20°C/min to heat the furnace from room temperature until the temperature reaches 950°C with a flow of an inert purified nitrogen gas at 100 mL/min. Then, the same heating rate was applied to increase the furnace temperature to 1300°C, and the gas being flowed at this combustion stage was changed to purified air. The trend of weight loss was recorded by thermogravimetry (TG) and derivative thermogravimetry (DTG). The data analysis was calculated based on weight loss procedure by the TGA software.

The Leco AC-350 bomb calorimeter was used to determine the calorific value (CV). The calorific value of a sample is determined by burning the sample in a controlled environment. The heat released by combustion is proportional to the calorific value of the substance. In the AC-350 bomb calorimeter, the weighed sample to be examined was located in a combustion vessel, which contains high-pressure atmospheric environment. The combustion vessel was surrounded by water and the sample is ignited. Succeeding that, the change in water temperature between pre-fire and post-fire was processed by the computer. The result was then being corrected for the length of fuse wire. The result, which is the calorific value, is then being shown on the display screen.

The Fourier-transform infrared (FTIR) spectra were recorded using a Perkin Elmer FTIR spectrophotometer. The spectral region from 4000 to 400 $cm^{-1}$ was investigated. This spectrometer has a spectral resolution of 0.5 $cm^{-1}$ standard with an accuracy in its higher wavelength of 0.01–3000 $cm^{-1}$. The resulting spectrum represents the sample absorption, following in its molecular fingerprint, due to its own functional groups. The FTIR spectra provide a quick qualitative technique that uses the standard IR spectra to identify the functional groups of the sample. The fundamental properties of the untreated PKS are summarized in **Table 1**.

### 2.5. Thermal decomposition using TGA

The thermal decomposition of the untreated and pretreated PKS was discovered by pyrolysis using a Mettler Toledo TGA/DSC 1 STAR$^e$ System. TGA provides a rapid method for determining the temperature-assisted decomposition profile of a sample and the kinetics of its thermal decomposition. A sample weight of 20 mg was inserted into 90 μL ceramic crucible. The pyrolysis temperature was raised from room temperature to 900°C. The experiments were conducted under heating rates of 10°C/min. The high-purity nitrogen with flow rate of 50 mL/min was used as a carrier gas to ensure the inert atmosphere during the pyrolysis

| **Properties** | **Value** |
|---|---|
| Elemental analysis (wt. %) | |
| Carbon | 47.67 |
| Hydrogen | 5.52 |
| Nitrogen | 0.39 |
| Sulphur | 0 |
| Oxygen[a] | 46.42 |
| Proximate analysis (wt. %) | |
| Moisture | 10.60 |
| Volatile matter | 77.54 |
| Fixed carbon | 10.95 |
| Ash | 0.91 |
| Calorific value (MJ/kg) | 18.20 |

[a]Calculated by different.

**Table 1.** Properties of untreated PKS.

process. The decomposition of the sample was analyzed using the TG curve, which showed the mass loss versus temperature and time curves of TGA experiment [34]. Also, the DTG curves, representing the rate of weight loss with the increasing temperature, indicated the determination of the decomposition and thermal characteristics of untreated and pretreated samples. Each untreated and pretreated sample was pyrolyzed at least twice. However, additional duplications were carried out where some inconsistencies were observed.

## 3. Results and discussion

### 3.1. Temperature profiles of torrefied PKS

**Figure 1** shows the temperature profiles of torrefied PKS at different processing time and microwave power levels. It shows that higher microwave power level contributed to increase the final temperature and heating rate. The torrefied PKS demonstrated increasing the final temperature and heating rate of 50.2–470.4°C and 12.6–117.6°C/min, respectively, when the microwave power level increased from 200 to 600 W in the first 4 min. These temperature profiles increased much steadily after about 4–8 min processing time. Conversely the temperature increment is not significant after 8–12 min processing time regardless the microwave power level. Therefore, higher processing time above 8 min was not necessary to upgrade the PKS. The microwave power level at 600 W with 4 min processing time was not suitable for upgrading the PKS, where it reached high heating rate of 117.5°C/min as the torrefaction requires heating rate equal or below 50°C/min [31].

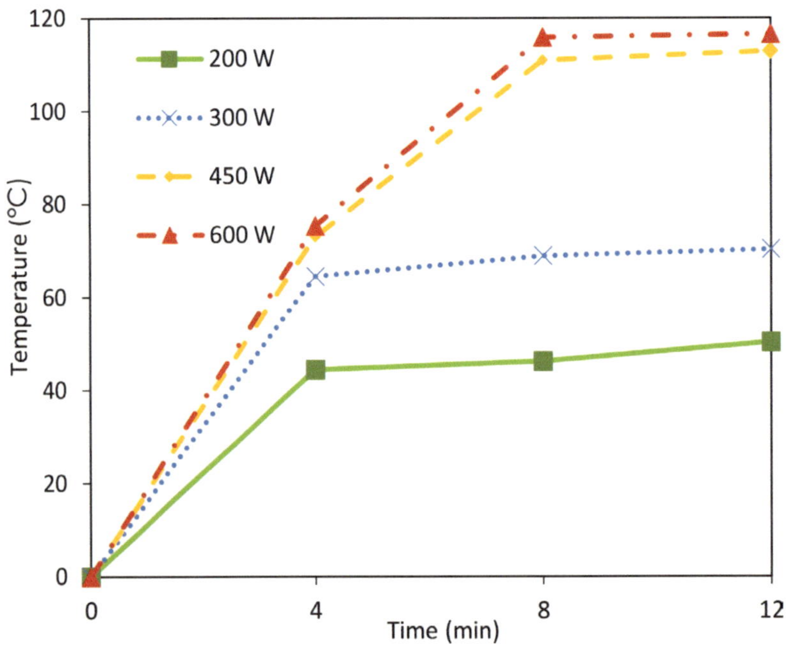


**Figure 1.** Temperature profile of torrefied PKS at various processing time and microwave power levels.

### 3.2. Mass and energy yield of torrefied PKS

**Figure 2(a)** and **(b)** represents the mass and energy yield of torrefied PKS, respectively. It can be seen that the microwave pretreatment decreased the mass and energy yield of torrefied PKS while applying higher microwave power at certain reaction time. For example, the mass yields of torrefied PKS for 4 min were 97.9, 96.1, 73.3, and 43.2%, while the energy yields were 100.5, 104.1, 84.9, and 52.8% at 200, 300, 450, and 600 W, respectively. The high mass and energy yield at 200 and 300 W were influenced from low reactivity at low microwave power level. While at moderate power level of 450 W, the mass and energy yield were reduced reasonably to 70.1 and 83.5%, respectively, at 8 min processing time. However, at a microwave power level of 600 W, the mass and energy yield extensively reduced toward 43.7 and 52.8%, respectively, at processing time of 4 min, because of the severe reaction at the high microwave power level. At the higher reaction temperatures, which also increase microwave power level, the volatilization reaction of biomass might become a predominant reaction during the pretreatment process. As a result, the mass and energy yield of biomass would be reduced. At operating condition of 450 W and 8 min, more than 70% of mass and 80% of energy have been remained in the torrefied PKS. This phenomenon should be due to the carbonization and volatilization reactions of biomass, which would take place at the same time during the pretreatment process.

### 3.3. Calorific value (CV) of torrefied PKS

The CV is one of the main parameters for fuels to be used in subsequent thermal conversion. **Figure 3** shows the CV of torrefied PKS at different microwave power levels for 8 min processing time. The CV of untreated PKS was 18.2 MJ/kg. At the microwave power level of 450 W, the torrefied PKS had the highest CV of 20.5 MJ/kg, which was 12.6% higher than untreated PKS. Commonly, higher microwave power level contributed to higher CV of pretreated feedstock. However, when the microwave power level increased from 450 to 600 W, the CV of

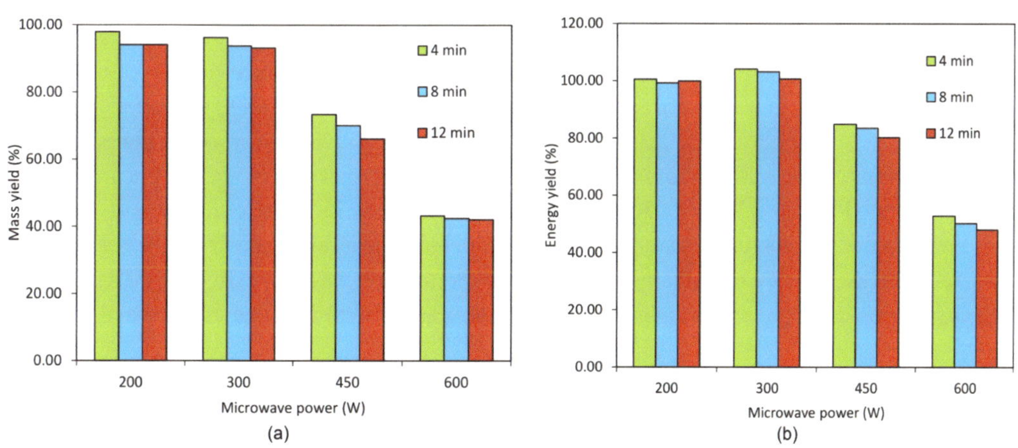


**Figure 2.** (a) Mass yield of torrefied PKS at various microwave power and processing time, (b) Energy yield of torrefied PKS at various microwave power and processing time.

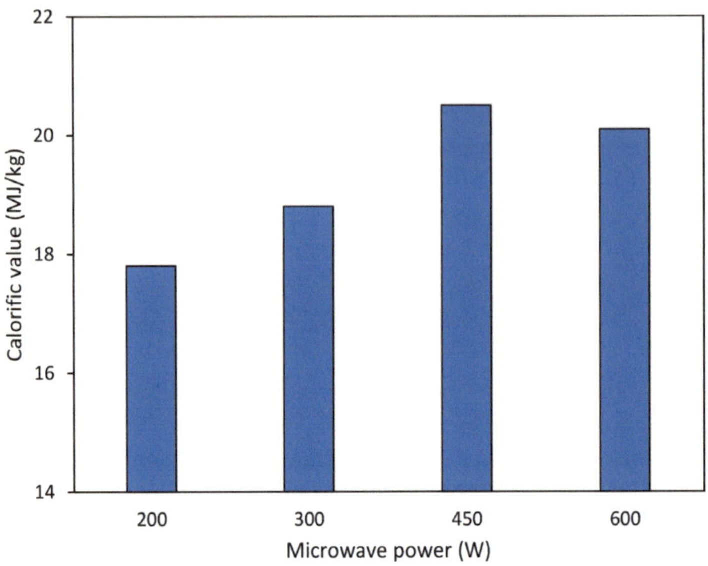


**Figure 3.** Calorific value of torrefied PKS at various microwave power with 8 min processing time.

torrefied PKS decreased. This may infer that when the reaction temperature is over 400°C due to higher microwave power levels (referring to **Figure 2(b)**), the fixed carbon content of biomass reduced resulting in the decrease of calorific value of torrefied PKS.

### 3.4. Proximate analysis of torrefied PKS

The torrefied PKS at 8 min processing time was chosen for proximate analysis. **Figure 4** shows the effect of microwave power level on moisture, volatile matter, and fixed carbon, respectively, of torrefied PKS. Generally, it can be seen that the moisture content and volatile matter decreased with increasing microwave power, in comparison to the untreated PKS. The results showed the characteristics of the torrefied PKS were altered due to high moisture content of untreated sample and its ability in absorbing microwave radiation. However, the fixed carbon of the torrefied PKS increased, with increasing microwave power level. The fixed carbon of the pretreated sample noticeably increased, representing a modification in quantity of energy per unit mass, which is related to the calorific value. Moreover, the decrease in volatile matter and moisture was observed. Since, the microwave pretreatment increases the carbon content, the fuel ratio of the irradiated samples eventually increased. This phenomenon was due to drying, volatilization, and decomposition of biomass feedstock during the pretreatment at higher microwave power.

The ratio of fixed carbon to volatile matter content, which is the fuel ratio, can indicate the accurate feedstock for thermal conversion. The fuel ratios of torrefied PKS at different microwave power levels with 8 min processing time are presented in **Figure 5**. After microwave pretreatment, the fuel ratios of pretreated materials significantly increased with increasing microwave power level. The fuel ratios of torrefied PKS increased from 0.48 to 2.85 when the microwave power levels increased from 200 to 600 W. The fuel ratio of 1.1 for pretreated PKS at 450 W is comparable with typical fuel ratio of bituminous coal, which is around 1.0–2.5 [35].

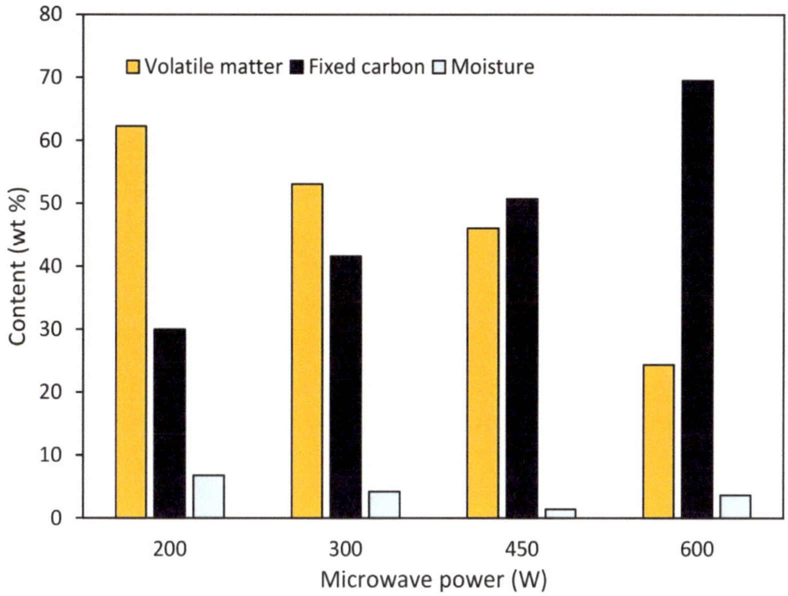


**Figure 4.** Volatile matter, fixed carbon and moisture content of torrefied PKS at various microwave power level with 8 min processing time.

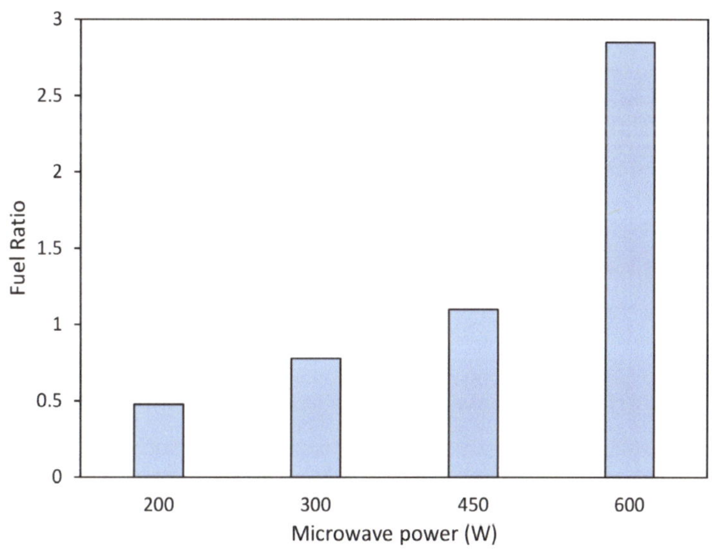


**Figure 5.** Fuel ratio of torrefied PKS at various microwave power with 8 min processing time.

### 3.5. Carbon and oxygen content of torrefied PKS

The torrefied PKS at 8 min processing time was chosen for ultimate analysis (carbon and oxygen content). **Figure 6** shows the effect of microwave power level on carbon and oxygen content of pretreated PKS. In general, the results indicate that oxygen decreased and carbon increased with the increase in microwave power level. The oxygen was reduced up to 43%

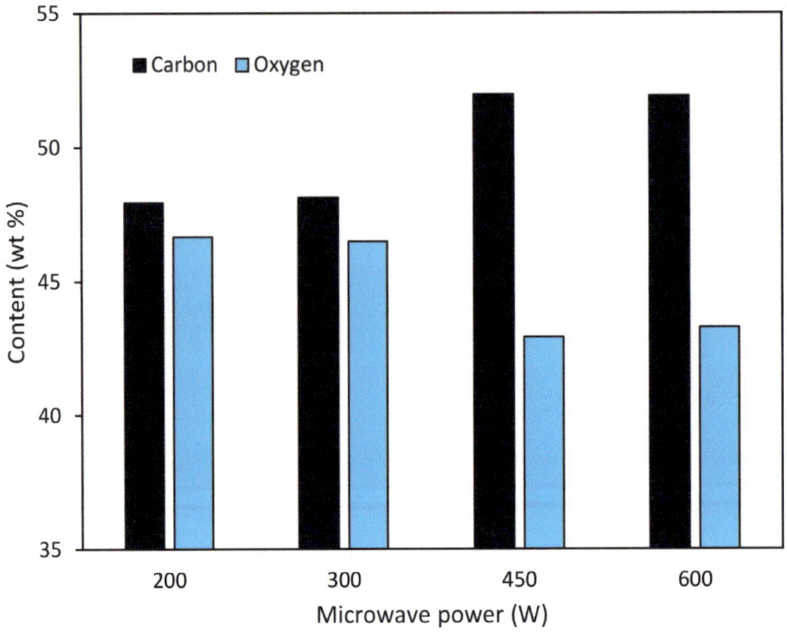


**Figure 6.** Carbon and oxygen content of torrefied PKS at various microwave power with 8 min processing time.

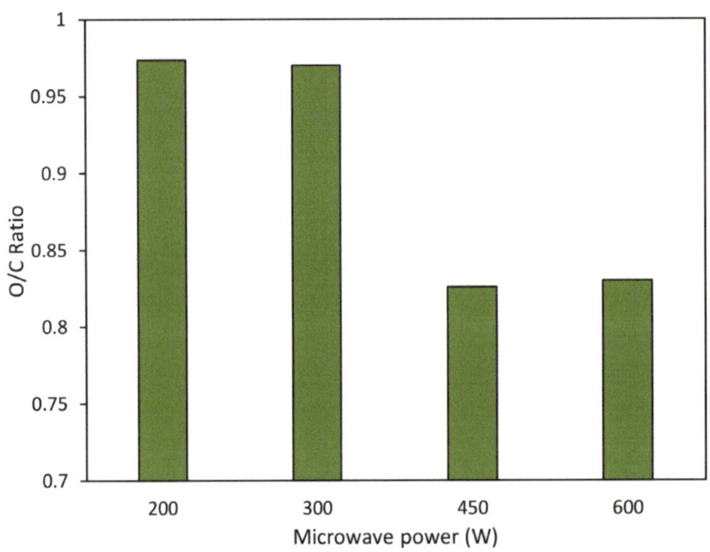


**Figure 7.** O/C ratio of torrefied PKS at various microwave power with 8 min processing time.

of pretreated PKS. On the contrary, carbon was increased up to 52 and 62% for torrefied PKS at the highest microwave power level of 600 W. The decrease in oxygen contents was generally attributable to the destruction of the hydroxyl group (–OH) in PKS during pretreatment, which consequently produced solid hydrophobic fuel. Eventually, by removing oxygen using microwave irradiation method, the energy density of the torrefied PKS increased.

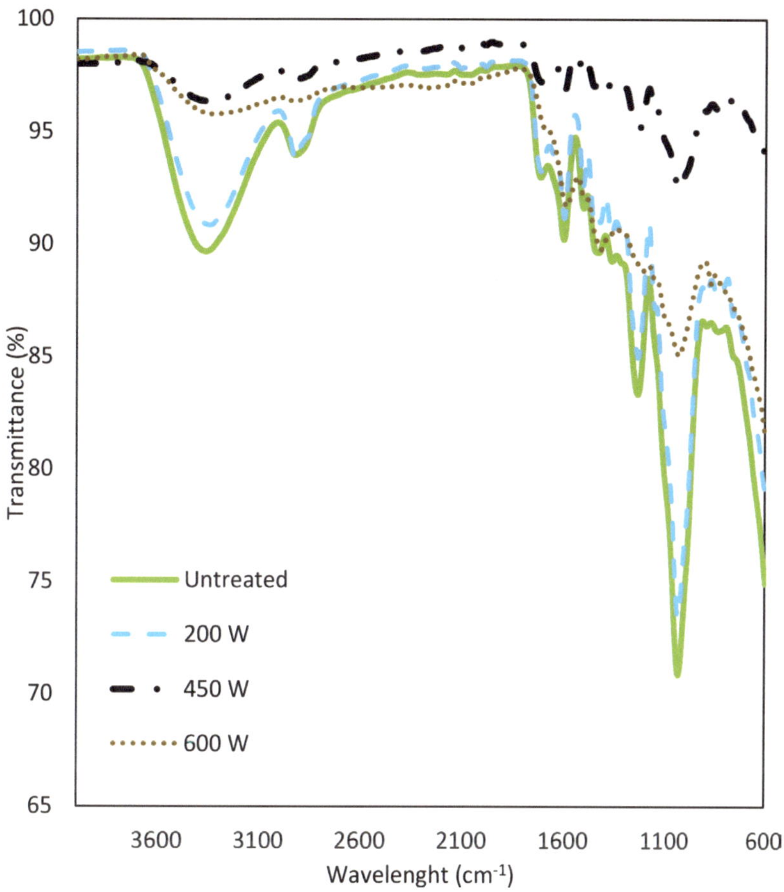


**Figure 8.** FTIR spectra of the untreated and pretreated PKS.

Due to the decomposition and elimination of volatile matter during pretreatment process, the oxygen mass fraction of the pretreated products will be lowered. Therefore, as illustrated in **Figure 7**, the O/C ratio of all torrefied samples was lower than that the untreated sample. As the microwave power level increased, the O/C ratio of torrefied PKS is gradually reduced as more volatile matter is being released as a result of the continuous decomposition process. The reduction of the atomic ratios also indicates the measures of conversion efficiency and oxidation degree of pretreated products [16].

### 3.6. Functional group analysis of torrefied PKS

The torrefied PKS at 8 min processing time with microwave power of 200, 300, 450, and 600 W were chosen for functional group analysis. The chemical structure difference of untreated and pretreated PKS at various microwave power level was characterized using FTIR as shown in **Figure 8**. The FTIR spectra of untreated and pretreated PKS are similar in shape, but the intensity of the peaks is different. A broad peak at 3400 $cm^{-1}$ for untreated PKS was associated to the hydroxyl group (–OH). These –OH groups exist with alcohols and phenols. The –OH peaks were remarkably decreased with the increase of the microwave power. The peaks at 2920 and 2880 $cm^{-1}$ indicated aliphatic methylene groups. The peak intensity of pretreated PKS was smaller than raw PKS at higher microwave power of 450 and 600 W. The carbonyl group (C=O) bonds were observed at 1750 $cm^{-1}$ corresponding to various acids, aldehydes, and ketones, which were formed by decomposition of cellulose and hemicellulose. The peak was smaller at higher torrefaction temperature, which was linked with breakdown of hemicellulose. Peaks at 1550 $cm^{-1}$ present alkenes of C=C stretching. The most concentrated peaks were observed in the range of 1500–1000 $cm^{-1}$ and assigned to C=O stretching and O–H deformation at organic compounds containing oxygen (alcohols, phenols, and ethers). Aromatic groups are represented by peak 790 $cm^{-1}$ for PKS.

### 3.7. Thermal decomposition of torrefied PKS

The torrefied PKS at 8 min processing time with microwave power of 300, 450, and 600 W were chosen for thermal decomposition in TGA. The analysis of pretreated sample at 200 W was not chosen because its characteristic was similar with the untreated sample as discussed in the earlier section. The DTG curve of untreated and pretreated PKS is presented in **Figure 9**. The untreated and pretreated PKS showed three noticeable peaks existed in the DTG curve.

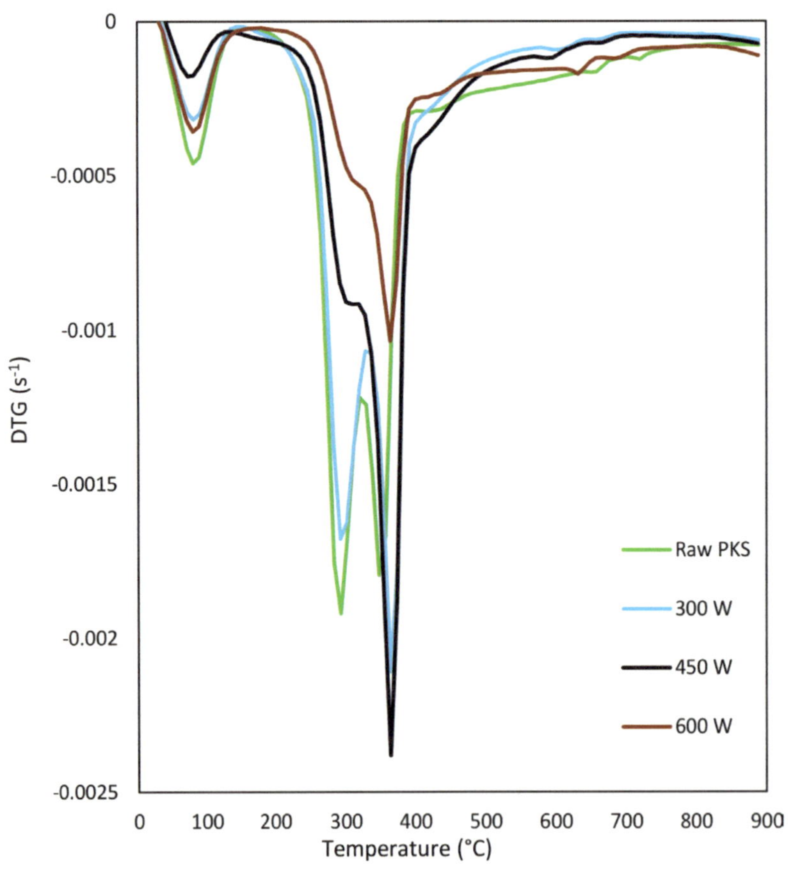


**Figure 9.** DTG curve of the untreated and pretreated PKS.

The main DTG peak at temperature range of 50–130°C indicated the moisture removal of the samples. The second DTG peak is located at 292 and 347°C, which referred to the maximum decomposition rate of hemicellulose and cellulose, respectively. After pretreated at microwave power level of 300 W, the peak was at 292°C and decreased slightly. However, the peak at 292°C moved to higher temperature at 373°C and significantly increased in the peak height. When the microwave power was increased to 450 and 600 W, the peak at 292 disappeared. Although the peak at 292°C moved to higher temperature at 373°C and significantly increased the peak height when pretreated at 450 W, there was a reduction in peak height after pretreated at 600 W. It is obvious that the second peak disappearance represents the hemicellulose lost at higher microwave power level at 450 W and above, whereas the third peak showed the cellulose retained, but the intensity was different. It is inferred that partial part of the cellulose and lignin remains and is not decomposed by the torrefaction using microwave irradiation [30].

## 4. Conclusions

This study presents the properties of torrefied PKS using thermal pretreatment via microwave irradiation. The torrefied PKS underwent physical and chemical modifications, which include mass reduction, rise in energy content, and change in chemical compositions. The increase in microwave power level showed the significant effect, which decreased the mass and energy yield of torrefied PKS. As the microwave power level increased, the moisture, volatile mater, oxygen content, and O/C ratio decreased. Among the microwave power-level variation studies, the carbon content and calorific value were enhanced to 55.91% and 21.20 MJ/kg, respectively, at microwave power of 450 W. The peak intensity of oxygenated functional group was reduced with the increase of the microwave power as presented in FTIR spectra. The TGA analysis has correlated the thermal decomposition with hemicellulose, cellulose, and lignin in torrefied PKS. The research can be concluded that the PKS can be upgraded via MI pretreatment to a value-added feedstock at microwave power of 450 W with processing time of 8 min. Thus, the torrefied PKS has the prospective to be applied in thermochemical conversion such pyrolysis, liquefaction, and gasification or co-conversion with coal.

## Acknowledgements

This research project is funded by the Ministry of Higher Education, Malaysia, under Fundamental Research Grant Scheme (FRGS/1/2017/TK10/UITM/02/11). The authors acknowledge Universiti Teknologi MARA and Universiti Malaysia Perlis for providing facilities during the research work.

## Conflict of interest

On behalf of all authors, the corresponding author states that there is no conflict of interest.

## Author details

Razi Ahmad[1,4*], Mohd Azlan Mohd Ishak[2,3], Nur Nasulhah Kasim[1] and Khudzir Ismail[1,3]

*Address all correspondence to: razi@unimap.edu.my

1 Faculty of Applied Sciences, Universiti Teknologi MARA, Shah Alam, Selangor, Malaysia

2 Faculty of Applied Sciences, Universiti Teknologi MARA, Perlis, Malaysia

3 Coal and Biomass Energy Research Group, Universiti Teknologi MARA, Shah Alam, Selangor, Malaysia

4 School of Environmental Engineering, Universiti Malaysia Perlis, Arau, Perlis, Malaysia

## References

[1] Emami Taba L, Irfan MF, Wan Daud WAM, Chakrabarti MH. The effect of temperature on various parameters in coal, biomass and CO-gasification: A review. Renewable and Sustainable Energy Reviews. 2012;**16**(8):5584-5596

[2] Heidenreich S, Foscolo PU. New concepts in biomass gasification. Progress in Energy and Combustion Science. 2015;**46**:72-95

[3] Mohr SH, Wang J, Ellem G, Ward J, Giurco D. Projection of world fossil fuels by country. Fuel. 2015;**141**:120-135

[4] Nhuchhen D, Basu P, Acharya B. A comprehensive review on biomass torrefaction. International Journal of Renewable Energy & Biofuels. 2014;**2014**:1-56

[5] Uemura Y, Omar WN, Tsutsui T, Yusup SB, Bt S. Torrefaction of oil palm wastes. Fuel. 2011;**90**(8):2585-2591

[6] Sabil KM, Aziz MA, Lal B, Uemura Y. Effects of torrefaction on the physiochemical properties of oil palm empty fruit bunches, mesocarp fiber and kernel shell. Biomass and Bioenergy. 2013;**56**:351-360

[7] Atnaw SM, Sulaiman SA, Yusup S. Syngas production from downdraft gasification of oil palm fronds. Energy. 2013;**61**:491-501

[8] Ahmad Kushairi D. Malaysian oil palm industry performance 2016 and prospects for 2017. In: Palm Oil Economic Review & Outlook Seminar 2017. 2017. p. 6

[9] Sabil KM, Aziz MA, Lal B, Uemura Y. Synthetic indicator on the severity of torrefaction of oil palm biomass residues through mass loss measurement. Applied Energy. 2013;**111**:821-826

[10] Brar JS, Singh K, Wang J, Kumar S. Cogasification of coal and biomass: A review. International Journal of Forestry Research. 2012;**2012**:1-10

[11] Ahmad R, Hamidin N, Ali UFM, Abidin CZA. Characterization of bio-oil from palm kernel shell pyrolysis. Journal of Mechanical Engineering Science. 2014;**7**(1):1134-1140

[12] Ahmad R, Ismail K, Ishak MAM, Kasim NN, Abidin CZA. Pretreatment of palm kernel shell by torrefaction for co-gasification. In: 4th IET Clean Energy and Technology Conference (CEAT 2016). 2016. pp. 1-6

[13] Chang ACC, Chang H-F, Lin F-J, Lin K-H, Chen C-H. Biomass gasification for hydrogen production. International Journal of Hydrogen Energy. 2011;**36**(21):14252-14260

[14] Howaniec N, Smoliński A. Effect of fuel blend composition on the efficiency of hydrogen-rich gas production in co-gasification of coal and biomass. Fuel. 2014;**128**:442-450

[15] Ahmad R, Azlan M, Ishak M, Kasim NN, Ismail K. Effect of different pretreatments on palm kernel shell and low-rank coal during co-gasification. Progress in Petrochemical Science. 2018;**2**(2):1-7

[16] Chen WH, Peng J, Bi XT. A state-of-the-art review of biomass torrefaction, densification and applications. Renewable and Sustainable Energy Reviews. 2015;**44**:847-866

[17] Asadullah M, Adi AM, Suhada N, Malek NH, Saringat MI, Azdarpour A. Optimization of palm kernel shell torrefaction to produce energy densified bio-coal. Energy Conversion and Management. 2014;**88**:1086-1093

[18] Nam H, Capareda S. Experimental investigation of torrefaction of two agricultural wastes of different composition using RSM (response surface methodology). Energy. 2015;**91**:507-516

[19] Satpathy SK, Tabil LG, Meda V, Naik SN, Prasad R. Torrefaction of wheat and barley straw after microwave heating. Fuel. 2014;**124**:269-278

[20] Mamvura TA, Pahla G, Muzenda E. Torrefaction of waste biomass for application in energy production in South Africa. South African Journal of Chemical Engineering. 2018;**25**:1-12

[21] Ibrahim RHH, Darvell LI, Jones JM, Williams A. Physicochemical characterisation of torrefied biomass. Journal of Analytical and Applied Pyrolysis. 2013;**103**:21-30

[22] Matali S, Rahman NA, Idris SS, Yaacob N, Alias AB. Lignocellulosic biomass solid fuel properties enhancement via torrefaction. Procedia Engineering. 2016;**148**:671-678

[23] Wang P, Howard BH. Impact of thermal pretreatment temperatures on woody biomass chemical composition, physical properties and microstructure. Energies. 2018;**11**(1):1-20

[24] Salema AA, Ani FN. Microwave-assisted pyrolysis of oil palm shell biomass using an overhead stirrer. Journal of Analytical and Applied Pyrolysis. 2012;**96**:162-172

[25] Motasemi F, Afzal MT. A review on the microwave-assisted pyrolysis technique. Renewable and Sustainable Energy Reviews. 2013;**28**:317-330

[26] Tahmasebi A, Yu J, Han Y, Li X. A study of chemical structure changes of Chinese lignite during fluidized-bed drying in nitrogen and air. Fuel Processing Technology. 2012;**101**:85-93

[27] Pickles CA, Gao F, Kelebek S. Microwave drying of a low-rank sub-bituminous coal. Minerals Engineering. 2014;**62**:31-42

[28] Kingman SW, Rowson NA. Microwave treatment of minerals—A review. Minerals Engineering. 1998;**11**(11):1081-1087

[29] Binner E, Lester E, Kingman S, Dodds C, Robinson J. A review of microwave coal processing. The Journal of Microwave Power and Electromagnetic Energy. 2014;**48**(1):35-60

[30] Wang MJ, Huang YF, Chiueh PT, Kuan WH, Lo SL. Microwave-induced torrefaction of rice husk and sugarcane residues. Energy. 2012;**37**(1):177-184

[31] Huang YF, Chen WR, Chiueh PT, Kuan WH, Lo SL. Microwave torrefaction of rice straw and pennisetum. Bioresource Technology. 2012;**123**:1-7

[32] Huang Y, Sung H, Chiueh P, Lo S. Microwave torrefaction of sewage sludge and leucaena. Journal of the Taiwan Institute of Chemical Engineers. 2017;**70**:236-243

[33] Sukiran MA, Abnisa F, Wan Daud WMA, Abu Bakar N, Loh SK. A review of torrefaction of oil palm solid wastes for biofuel production. Energy Conversion and Management. 2017;**149**:101-120

[34] Vuthaluru HB. Investigations into the pyrolytic behaviour of coal/biomass blends using thermogravimetric analysis. Bioresource Technology. 2004;**92**:187-195

[35] Sutcu H. Pyrolysis by thermogravimetric analysis of blends of peat with coals of different characteristics and biomass. Journal of the Chinese Institute of Chemical Engineers. 2007;**38**(3-4):245-249

Chapter 4

# Chemical Kinetic and High Fidelity Modeling of Transesterification

Isam Janajreh and Manar Almazrouei

Additional information is available at the end of the chapter

http://dx.doi.org/10.5772/intechopen.80008

**Abstract**

The modeling and simulation of transesterification require an understanding of the chemical reactions that take place inside the reactor. The development of reaction mechanism of the multiple step triglyceride, triglycerides and mono-glycerides and their reversal reaction is beyond the interest of chemical or mechanical engineers, whose main interests are to assess the conversion overall and to establish performance process metrics. This chapter undertakes the transesterification conversion by firstly establishing and formulating the overall process kinetics as far as the rate constant and activation energy. Secondly, use the obtained kinetic values to carry out high fidelity reactive flow of the multiple species which are co-present inside the reactor and otherwise complex to capture experimentally. Following these two steps, this work provides qualitative and quantitative information on the concentration of the reactants, intermediates and the overall yield. This two-step-approach can also be utilized as reactor design tool and gaining in-depth insight on reaction progress and species distribution. Experimental results, high-fidelity numerical results, and parametric sensitivity studies will be introduced and discussed.

**Keywords:** chemical kinetic, transesterification, CFD, biodiesel, crude glycerol

## 1. Introduction

Stoichiometrically and theoretically speaking, transesterification consumes 1 mole of triglyceride and 3 moles of alcohol to produce 3 moles of fatty acid methyl esters (FAME) and 1 mole of crude glycerol. Practically, unconverted triglyceride (TG) and intermediates (i.e. diglyceride (DG) and monoglyceride (MG)) co-present in the yield which signifies the incompletion of the reaction [1]. As these reactions are mildly influenced by temperature and pressure because of their nearly equal heat of formation and liquid phase, the increase in the molarity of the alcohol

| Species | Chemical formula | Molecular weight (g/mol) | Viscosity (kg/m.s) | $C_P$ (J/kg.°C) | Density (kg/m$^3$) |
|---|---|---|---|---|---|
| Methanol | $CH_4O$ | 32 | 3.96E-4 | 1.470E3 | 791.8 |
| Waste oil or Triglyceride | $C_{54}H_{105}O_6$ | 848 | 1.61E-2 | 2.2E3 | 883.3 |
| Diglyceride | $C_{37}H_{72}O_5$ | 596 | — | — | 880 |
| Monoglyceride | $C_{20}H_{40}O_4$ | 344 | — | — | 875 |
| Biodiesel | $C_{18}H_{36}O_6$ | 284 | 1.12E-3 | 1.187E3 | 870 |
| Glycerol | $C_3H_9O_3$ | 93 | 1.412E0 | 0238.6 | 1261 |

**Table 1.** Summary of species properties and molecular weight (MW) [5].

promotes the desired forward reaction [2]. Contrary to well-known hydrocarbon fuels that are characterized by fixed thermodynamic and physical properties, the TG, DG, MG have no fixed chemical formula and neither their thermodynamic properties, such as standard enthalpy or specific heats, nor physical ones like density or viscosity, are consistent throughout the literature [3, 4]. Therefore, material characterization is an essential step in the modeling of the transesterification process. The extent of these properties depends on the complexity and comprehensiveness of the simulation, from a simple incompressible flow that requires only viscosity and density, to a complex non-isothermal flow that requires heat of formation, specific heat, thermal conductivity and their associated diffusions. However, as these properties can be derived following the American Society for Testing and Materials (ASTM) standards, their reactions are more complex. **Table 1** summarizes some of the utilized properties for the waste oil, TG, DG, and MG used in the work of Noureddine and Zhu who were amongst the pioneers of quantifying transesterification reaction kinetics [5].

Setting up a reaction mechanism of numerous species or elements and hundreds of reactions, while accounting for reaction radicals, is rather impractical for engineers. The overall reaction can be captured through well controlled conditions and yield assessment procedures that can save the pain of the development or use these reaction mechanisms. This chapter undertakes the conventional transesterification at different process temperatures, highlighting their influence on the yield and their distribution inside the reactor.

## 2. Biodiesel feedstock and the reactor device

While corn, sunflower, and palm oil are readily available in the market, their utilization may raise a strong debate on land for food vs. land for energy. To rule out this debate, waste cooking oil (WCO) is used instead, because it is abundantly available feedstock, inexpensive and most of the time its disposal into sewerage systems is associated with environmental concerns. The supply chain of collecting this abundant source is beyond the scope of this chapter. However, it is important to state that residential communities are in general in favor of trapping this problematic sewerage source that is responsible for clogging the plumbing systems. Collection also can be facilitated by using drum type or smaller plastic containers provided to the local restaurant, school/university canteens, residential communities as a privately owned small business, or through municipality.

The collected WCO generally requires pretreatment that can be facilitated at a moderate temperature to maintain its liquid and less viscous form. One can use the heat of the summer (45–50°C) at which unsaturated and saturated steric fatty acid stays in liquid form, and 10–20 μm filtration can be used to eliminate any suspended oil solid residuals. Dehydrating of the WCO is also required, during which any water content brought by the processed food into the waste cooking oil is liberated through evaporation. At the laboratory scale, stirring and heating pad at near 110°C for several hours can perfect this task. Process methanol and catalyst NaOH or KOH can also be substituted with commercially available grade instead of high purity pharmaceutical grade/Sigma-Aldrich that can also leverage process economically. Once pretreatment of the feedstock is done, the NaOH solid catalyst in the form of small ballets is dissolved into methanol at the stipulated ratio, i.e. 0.5–1% by mass of WCO. This process can be facilitated under moderate heating and a temperature below 60°C and stirring forming the meth oxide reactant solution. In the lab, multiple transesterification reaction experiments can be conducted simultaneously under the same temperature and stirring rate to reduce experimental sequence and human error. This can be carried out using a multiple dissolution apparatus such as those provided by *Agilent Technologies,* featuring 6–12 reactor vessels as depicted in **Figure 1** [6]. The caps are tightly fitted and are equipped with direct access ports for sampling without process interruption. They all are also set on thermally controlled wells. The process in these individual small-scale batch reactors resemble those carried out in larger

**Figure 1.** Dissolution apparatus representing eight multiple 1-L batch reactors and *HomeBiodeselKits* 500-liter batch reactor.

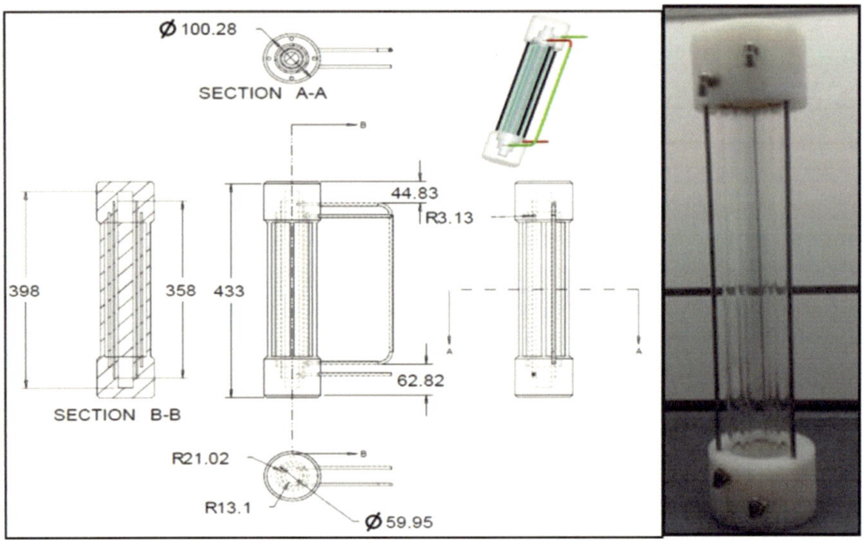


**Figure 2.** Patented tubular continuous biodiesel reactor 1–8L/hr [19].

scale ones provided by *HomeBiodeselKits*, as depicted in **Figure 1**. Beside batch method, a patented continuous reactor is also depicted in **Figure 2** featuring compactness, continuity, and scalability. This reactor reported better efficacy observed by the higher throughput and better quality of product than the batch reactor.

The continuous reactor consists of a tubular reactor of two or more upright concentric cylinders. Each is equipped with inflow at the bottom and outflow at the top port, connected by 5-mm ϕ chemically resistive hoses. The inflow and outflow are configured circumferentially in each cylinder, rendering the flow more residence time due to their helical trajectory. The details of the reactor's dimensions and geometry can be found in the work of Janajreh et al. [7]. A peristaltic or a diaphragm pump type is used to inject the two reactants, i.e. pretreated oil and methoxide into the reactor at the stipulated molar ratio. The mass flow rate is typically being set beyond laminar limits for turbulence mixing that help the transesterification reaction. The mixture from the batch dissolution is periodically recovered in 10 mm vials, and is refrigerated to halt their progressive reactions for downstream species analysis. The analysis of the compositions of TG, DG, MG, FAME, glycerol (GL) and alcohol (AL) is carried out using a standard Gas Chromatography Mass Spectrometry (GC/MS) equipment as those provided by *Thermo Scientific DSQ II* equipped with a flame ionization detector.

In the GC/MS analytical equipment, the FAME column is initially calibrated using standard biodiesel and glycerol samples to ensure precise qualitative and quantitative analysis. The collected and refrigerated vials are obtained at numerous and progressive reaction intervals of 15-time steps over 2 hr of reaction time [8]. The breakdown of sample species composition is acquired using methods similar to those carried out by Noureddini and Zhu [5] and Janajreh et al. [7]. Transesterification is performed at 6:1 alcohol to oil molar ratio and 0.5% NaOH by mass of WCO, and both at different temperatures of 50 and 60°C.

## 3. Transesterification method

**Figure 3** shows the three-step details of the three equilibrium reactions process of transesterification reaction as indicated by Noureddine et al. [5] and Janajreh et al. [9].

This equilibrium given in **Figure 1** is represented by six coupled first order differential equations including the overall shunt reaction as per Eqs. (1)–(6) [10] and these are written as:

$$d[TG]/dt = -k_1[TG][AL] + k_2[E][DG] - k_7[TG][AL]^3 + k_8[E]^3[GL] \quad (1)$$

$$d[DG]/dt = -k_3[DG][AL] + k_4[E][MG] + k_1[TG][AL] - k_2[E][DG] \quad (2)$$

$$d[MG]/dt = -k_5[MG][AL] + k_6[E][GL] + k_3[DG][AL] - k_4[E][MG] \quad (3)$$

$$d[E]/dt = k_1[TG][AL] - k_2[E][DG] + k_3[DG][AL] - k_4[E][MG] + k_5[MG][AL] - k_6[E][GL] + k_7[TG][AL]^3 - k_8[E]^3[GL] \quad (4)$$

$$d[GL]/dt = k_5[MG][AL] - k_6[E][GL] + k_7[TG][AL]^3 - k_8[E]^3[GL] \quad (5)$$

$$d[AL]/dt = -d[E]/dt \quad (6)$$

In these equations, *k* is the rate constant of each of the forward and backward reaction of **Figure 3** and (*x*) is the concentration of species *x*, the *TG*, *DG*, *MG*, *AL*, *E*, and *GL* are

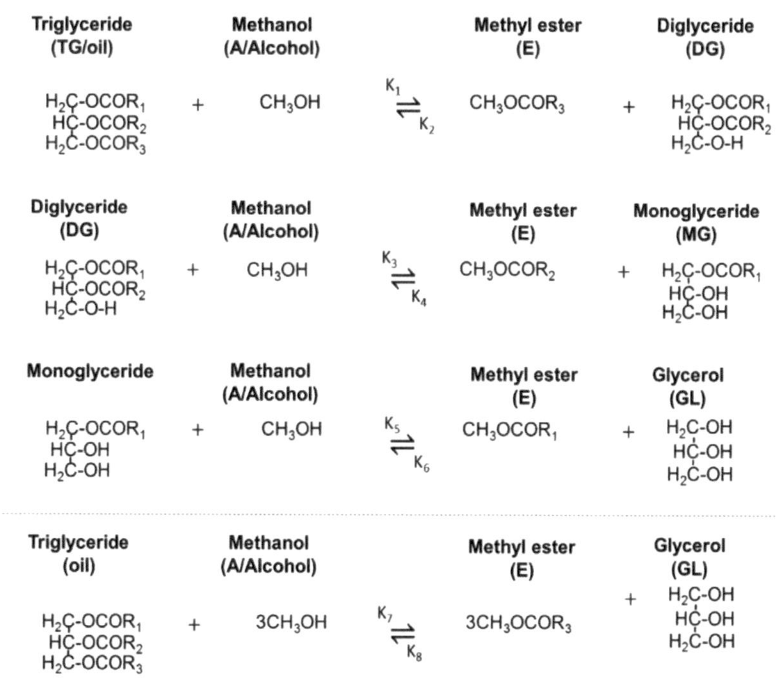


**Figure 3.** Transesterification reaction mechanism.

respectively the Triglyceride, Diglyceride, Monoglyceride, Alcohol, Fatty Acid Methyl Ester and Glycerol. This system of PDEs Eqs. (1)–(6) is solved for $k_1$ through $k_8$ at minimum root mean square following the temporal measurements of each specie and using Eq. (7) as:

$$Ax = b \ or \ x = A^{-1}b \tag{7}$$

where $A$ is the coefficient matrix of the species concentration at each time step obtained through GC/MS analysis, $x$ is the evaluated rate constant vector ($k_1$ through $k_8$) and $b$ is obtained by evaluating the time derivative of the measured concentrations. The rate constant ($k$) of each reaction is expressed by the Arrhenius Eq. (8) [11] as follows:

$$k = A_o e^{-E/RT} \text{ or } \ln k = \ln A_o - \frac{E}{R}\frac{1}{T} \tag{8}$$

where $A_o$ ($s^{-1}$) is the pre-exponential constant, $E$ (J/mol) is the activation energy corresponding to each of the eight reactions, $R$ (J/mol K) is the molar universal gas constant and $T$ is the process temperature expressed in degree Kelvin. Therefore, from the slope and intercept of the natural log line of $K$ and $1/T$ one can infer the activation energy (which is normalized by the gas constant) and the natural log of the pre-constant, respectively. Following previous work on chemical kinetics [5, 12], the reaction rate constants Eqs. (1)–(8) are solved simultaneously using MATLAB which enables direct inference of $E$ and $A_o$ [13].

## 4. Transesterification kinetics evaluation

Transesterification is a slowly reversible reaction, where only the forward reactions are desired for the production of biodiesel. It is easy to say that lower activation energies are favored for forward reactions (*E1*, *E3*, *E5*, *E7*) while higher activation energies are preferred for backward reactions (*E2*, *E4*, *E6*, *E8*). Unfortunately, this is not as simple as will be stated below. The obtained GC/MS results for this conventional transesterification at 0.5% NaOH, 6:1 molar ratio and temperature of 50 and 60°C, in terms of mole fraction, are depicted in **Figure 4**. It shows the rates of consumption of TG and Alcohol into the formation of methylesters (*E*) and *GL* as well as the intermediate species, i.e. *DG* and *MG*.

The *E* production and its byproduct *GL* demonstrate a sigmoidal trend: a slow delay in the reaction rate at the start is then followed by a quick surge in the production, and a final slow rate near the reaction completion. This trend characterizes the changing mechanisms reaction that swings from the slow mass transfer controlled to the fast kinetically controlled and back to slow reactivity as equilibrium is attained [5]. This can be explained by the poor diffusion between the two phase reactants system, i.e. *TG* and *AL*. As soon as the methyl ester is formed the solubility of the two phases into one single phase is increased drastically which causes the production surge. This is stipulated by any sample taken at the first minute of the reaction, which tends to form two separate phases. Samples taken thereafter appear to be a more homogenized single phase as sampling time progress. At the onset of the formation of the single phase, mixing becomes less important and hence temperature can play more important

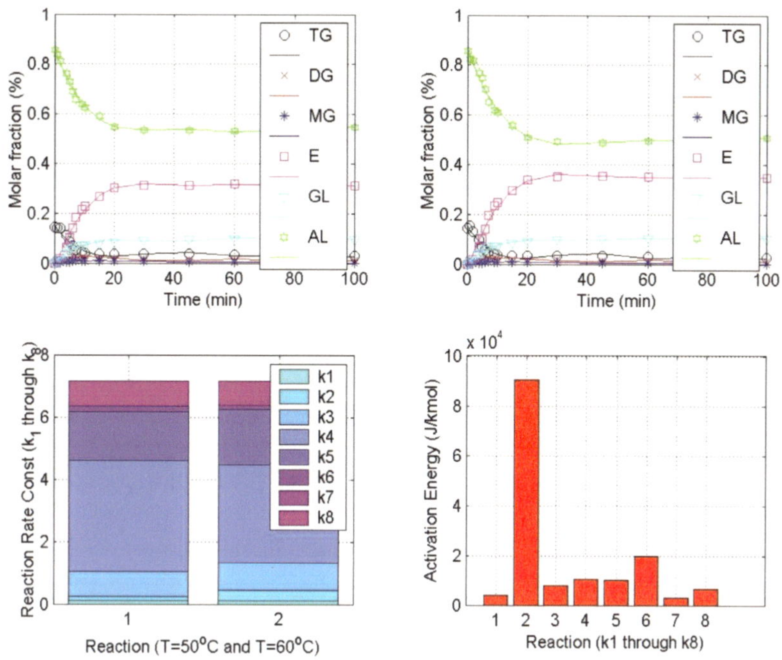


**Figure 4.** Transesterification of WCO conversion results at 50°C (top left) and 60°C (top right). Corresponding chemical kinetic reaction rate constants (bottom left) and activation energy (bottom right).

role to kinetically propel these reactions. In general, an increase in temperature provides a higher energy state for the molecules leading to more collisions and also improves the solubility of the reactants. As these reactions are reversible, the consumption of any of the reactants must be avoided to steer and ensure these reactions of **Figure 3** forward.

As can be shown in **Figure 4** of the GC/MS results obtained at the two temperatures, *DG* and *MG* are intermediate species. They appear following mass transfer limitation for a relatively shorter time, before they asymptotically vanish. The amount of unreacted alcohol at T = 60°C is shown to be lower than those obtained at T = 50°C, as per **Figure 4**. On the contrary, the *FAME* production and its byproduct *GL* is noticeably higher at higher temperature (T = 60°C) than the lower process temperature (T = 50°C). The rate constants for the two process temperatures are depicted side by side in the bottom left 'staked bar plot' of **Figure 4**. These values are relatively comparable, except for $k_2$ and $k_4$. Considering the lower temperature stacked bar (left side), it shows the low value of $k_1$ and $k_2$ reaction constants compared to the rest of reaction constants placing them as the rate limiting reactions. As soon as the product of these reactions becomes available, higher rates for their intermediate yield takeover. Comparing the lower and higher temperature $k_s$, the higher temperature results in higher $k_1$ and $k_2$, which are the rate limiting reactions, while the rest are comparable. This explains the higher initial yield beyond the two-phase mass transfer limitation that lasts within the first minutes of the reaction. These evaluated activation energies are also depicted in the bar plot of **Figure 4** on the bottom right. These

| Designated K or E | Rate constant at T = 50°C | Rate constant at T = 60°C | Activation energy (J/mol) | Pre-constant ($sec^{-1}$) |
|---|---|---|---|---|
| TG + AL—E + DG | 0.1213 | 0.1157 | 4103 | 77,983,755.8 |
| E + DG—TG + AL | 0.1167 | 0.3276 | 90,540 | 449,020.085 |
| DG + AL—E + MG | 0.8177 | 0.8947 | 7896 | 1.0976 E12 |
| E + MG—DG + AL | 3.5379 | 3.1354 | 10,597 | 4570,1701.3 |
| MG + AL—E + GL | 1.5943 | 1.7922 | 10,267 | 767.86 |
| GL + E—MG + AL | 0.1661 | 0.1322 | 19,980 | 32,870.58 |
| TG + 3AL—3E + GL | 0.001 | 0.001 | 3089 | 77,983,755.8 |
| GL + 3E—TG + 3A | 0.8149 | 0.7568 | 6490 | 449,020.085 |

**Table 2.** Summary of the rate constants values of the reactions at two temperatures and their activation energy.

are consistent with those reported experimentally by Feedman and coworkers [14, 15]. Based on the Arrhenius equations, they reported activation energies in the transesterification of the soybean oil in the range of 33,472–83,400 J/mol [16, 17]. These values are tabulated in **Table 2**, along with the rate constant. The activation energy is the highest for the reverse of the first reaction, DG + E—TG + AL suggesting this reaction is not as spontaneous as the others and reverse formation of TG is rather less expected. This is due to surplus of the Methanol alcohol that forces the forward reactivity of the TG + AL—DG + E reaction. This is also evident from all the forward reactions which are at lower activation energy than their associated reversal reaction.

The evaluated kinetics will be utilized in the development of a high fidelity reaction model. The model will be based on reactive flow of the six species, TG, AL, DG, MG, E and GL and their eight reactions. These are types of volume homogenous reactions. The considered reactor will be a tube type subjected to appropriate conditions and enabling the assessment of the yield and species distribution.

In summary, the required kinetics that governs the reactions and change of species has been evaluated. Results are intuitively correct, as a higher temperature results in higher rate constants. These data will be used in the high fidelity reactive flow analysis that will be detailed next. It is important to emphasize that importance of these simulations in capturing the overall species and their distribution in the reaction device. This is becoming the tool for the development of the new and innovative reactors. The observations procured from performing the chemical kinetic study enables the shift in conventional transesterification towards a modified sonication kinetic data for further reactive flow analysis. Comparison between regular mixing and ultra- sound assisted procedures proves the supremacy of the latter in terms of fastening the forward reactions by increasing reaction rates and decreasing activation energies of forward reactions. Note that the comparative study is performed to obtain an idea of the trend of sonication and not to get exact chemical kinetics.

# 5. High-fidelity transesterification model development

In order to sharpen our understanding of the transesterification reaction progression and species distribution, a high fidelity reactive flow model for transesterification is developed. The basis of the model is computational fluid dynamics (CFD). Several research software have adopted CFD environment to solve multiple physics, whereby some are open sources and others are more commercially tuned. The transesterification model is developed within the finite volume CFD based software of Ansys/Fluent 17.1 [18].

## 5.1. Governing equations

We here focus on the modeling of the tubular reactor that depicted earlier in **Figure 2** which has numerous advantages over the batch reactor, i.e. continuous, compactness, better yield [19]. Modeling involves the application of flow continuity, momentum, and energy equations. Furthermore, the flow is characterized as a mixture of multiple reacting species, incompressible, viscous, and turbulent. The reaction is assumed to start as soon as the reactant components are met inside the reactor. The flow is governed by the Navier-Stokes equation, which is associated with temporal, advective, viscous, and any source term and is written as:

$$\frac{\partial}{\partial t}(\varnothing) + \frac{\partial}{\partial x_i}(u_i\varnothing) = \frac{-\partial}{\partial x_i}\left(\Gamma_\varnothing \frac{\partial \varnothing}{\partial x_i}\right) + S_\varnothing \tag{9}$$

where $u_i$ is the velocity and $S_\varnothing$ is the source term due to the dispersed phase interaction. The $\varnothing$ is the flow dependent variable and when represents the density ($\rho$), velocity density multiple ($\rho u_i$) and the temperature ($T$) it yields the continuity, the momentum, and the energy equations, respectively. It takes also the turbulent scalars, i.e. kinetic energy ($k$) and dissipation rate ($\varepsilon$). These two equations in steady state are expressed as:

$$\begin{aligned} \rho u_i \frac{\partial k}{\partial x_i} &= \mu_t\left(\frac{\partial u_j}{\partial x_i} + \frac{\partial u_i}{\partial x_j}\right)\frac{\partial u_j}{\partial x_i} + -\frac{\partial}{\partial x_i}\left(\frac{\mu_t}{\sigma_k}\frac{\partial k}{\partial x_i}\right) - \rho\varepsilon \\ \rho u_i \frac{\partial \varepsilon}{\partial x_i} &= C_{1\varepsilon}\frac{\mu_t \varepsilon}{k}\left(\frac{\partial u_j}{\partial x_i} + \frac{\partial u_i}{\partial x_j}\right)\frac{\partial u_j}{\partial x_i} + \frac{\partial}{\partial x_i}\left(\frac{\mu_t}{\sigma_\varepsilon}\frac{\partial \varepsilon}{\partial x_i}\right) - C_{2\varepsilon}\frac{\rho\varepsilon^2}{k} \end{aligned} \tag{10}$$

The left terms govern the advective while the right hand terms govern respectively the generation, the diffusion, and destruction of the turbulent quantities. In these equations, $\mu_t$ is the eddy viscosity parameter and it overwhelms the laminar viscosity and is written as:

$$\mu_t = f_\mu C_\mu \rho k^2 / \varepsilon \tag{11}$$

where $f_\mu$ and $C_\mu$ are flow dependent constants and $C_{1\varepsilon}$, $C_{2\varepsilon}$,$\sigma_k$ and $\sigma_s$ are tuning empirical constants. The transport equations that govern the of species $m_i$ take the following form:

$$\frac{\partial}{\partial t}(\rho m_i) + \frac{\partial}{\partial x_j}(\rho u_j m_i) = \frac{\partial}{\partial x_j}(\rho D_{i,m} + \mu_t / Sc_t)\frac{\partial m_i}{\partial x_j} + R_i + S_i \tag{12}$$

Here $D_{i,m}$ is the diffusion coefficient of $i$ specie within the $m$ bulk species and $Sc_t$ is the turbulent Schmidt number defined as the ratio of the eddy viscosity t to the eddy diffusivity $D_{i,m}$. Eq. 10 incorporate an additional source term $R_i$ which accounts for species reaction. This term is governed by the reaction stoichiometry of the specie $m_i$ and is written as:

$$\sum_{i=1}^{N} v'_{i,r} S_i \underset{k_{b,r}}{\overset{k_{f,r}}{\Leftrightarrow}} \sum_{i=1}^{N} v''_{i,r} S_i \tag{13}$$

The reaction rate is proportional to the concentration of the reaction species (both reactants and products) to an ordered of specified power coefficients that can be written as:

$$R_{i,r} = M_{i,r} (v''_{i,r} - v'_{i,r}) \left( k_f \prod_{j=1}^{N} C_{j,r}^{\eta^{\bullet}_{rj}} - k_b \prod_{j=1}^{N} C_{j,r}^{v'^{\bullet}_{j,r}} \right) \tag{14}$$

where $k_f$ $(k_1, k_3, k_5, k_7)$ and $k_b$ $(k_2, k_4, k_6, k_8)$ are the reaction rate constants evaluated according to Eq. (8) for the forward and backward of the above Arrhenius equation Eq. (13) that summarizes equations of **Figure 3**. The $Cj$ are the molar concentration of jth specie to the order of its stoichiometric coefficient $v$ while $\eta$ is the reaction order. The $M_i$ is the molecular weight of species i [20]. It is a good practice to pursue CFD as a non-reactive flow first before accounting the influence of reactions that numerically may induce earlier instability for the solution. After achieving near steady and consistent solution and influence of reaction can be included and analyzed.

### 5.2. CFD setup and mesh sensitivity analysis

The model of the considered tubular reactor is shown below in **Figure 5** along with its dimensions as summarized in **Table 3**.

A hybrid mesh of hexagonal and pyramid type is used to maintain the size within the processing capacity of the current laptop for an engineer. The mesh is established in *ANSYS Design Modeler,* consisting of two concentric tubular chambers along with the connecting tubing as shown in **Figure 6**. A baseline and another two refined meshes were created to study the dependency and the goodness of the mesh influence on the solution. The properties of the reactants and products are summarized in **Table 4** as per the work of Narvaez et al. [21]. The ideal mixture is used to determine the mixture properties from the known properties of the individual specie as per Eq. (15).

$$\varphi_m = \sum_{i=1}^{n} m_{f,i} \varphi_i \tag{15}$$

The boundary conditions are assigned as flow, constant velocity at the inlet and isothermal fluid. The boundary at the top outlet was defined as fixed pressure outlet at 0-Pa gauge pressure while the boundary walls are subjected to zero velocity, i.e. no slip and no penetration. The turbulence is accounted for via standard $k$-$\varepsilon$ mode and model constants were kept at their common values [22–24]. The setup was initialized at the same values of velocity from the

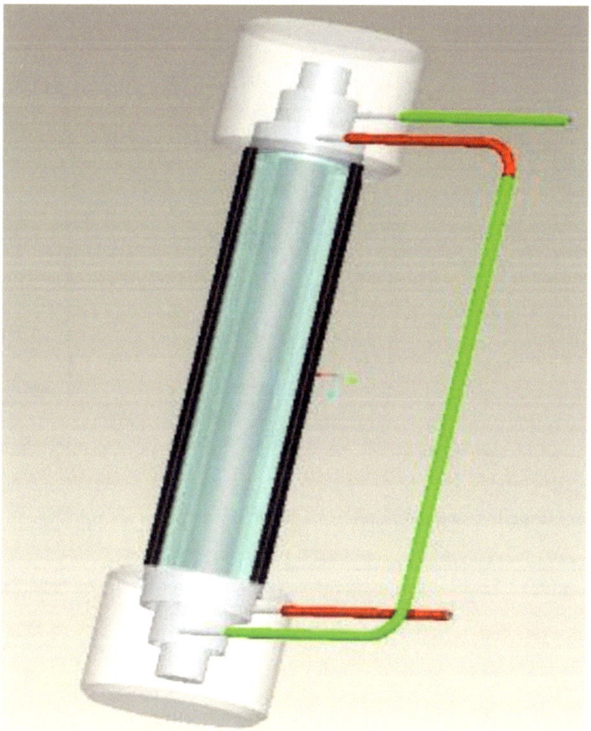

Figure 5. Tubular reactor model.

| Component | Outer cylinder (glass) | Middle cylinder (metal) | Inner cylinder (metal) |
|---|---|---|---|
| Outer diameter (mm) | 59.9 | 42.0 | 26.2 |
| Inner diameter (mm) | 51.9 | 36.7 | 21.8 |
| Thickness (mm) | 4.0 | 2.7 | 2.2 |
| Length (mm) | 272 | 313 | 353 |

Table 3. Dimensions of the reactor geometry components.

inlet boundary condition. The material initially is assumed as an ideal mixture that has the viscosity and density property proportion of 6:1 methanol-to-waste cooking oil mole fraction.

Results of the unreacted isothermal flow distribution within the reactor, or cold flow as referred to in some literature, is depicted in **Figure 7**. Velocity vector colored by the residence time is used for the three level meshes. Results of residence time parameter is used as the most pronounced parameter for the intended reactive fluid to assess the mesh accuracy. These results are summarized in **Table 5** for the three levels of the mesh for comparison. The baseline residence time is within 2% deviation from the next refined mesh, hence it is used for the rest of the analysis. However, the coarse mesh showed a sustainable deviation of nearly 10% that fails to be qualified as an accurate and representative mesh.

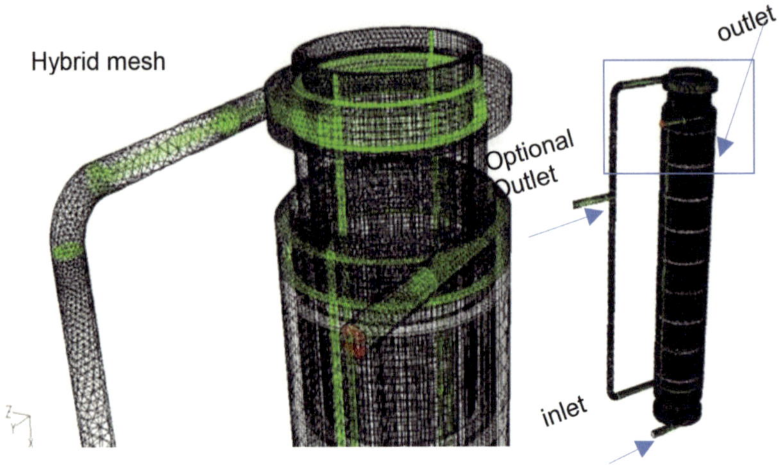


**Figure 6.** Discretized reactor domain using hybrid 3-D baseline mesh.

| Species | Chemical formula | Molecular weight (g/mol) | Viscosity (kg/m.s) | Cp (J/kg.°C) | Density (kg/m$^3$) |
|---|---|---|---|---|---|
| Methanol | $CH_4O$ | 32 | 3.96e-4 | 1.470e3 | 791.8 |
| Waste oil | $C_{54}H_{104}O_6$ | 849 | 1.61e-2 | 2.2e3 | 883.3 |
| Biodiesel | $C_{18}H_{36}O_2$ | 284 | 1.12e-3 | 1.187e3 | 870 |
| Glycerol | $C_3H_8O_3$ | 92 | 1.412e0 | 0238.6 | 1261 |

**Table 4.** Properties of the flow species.

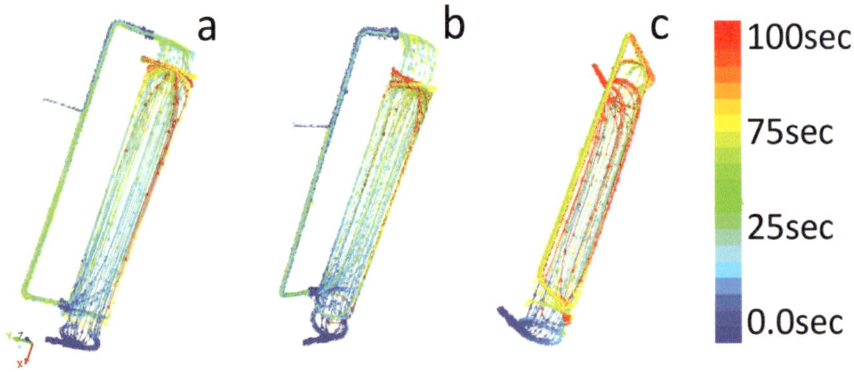


**Figure 7.** Flow trajectory colored by the resident time for the (a) baseline, (b) fine and (c) very fine.

A total resident time of the order of $10^3$ seconds is observed in each of the three meshes. If the flow was injected at relatively high velocity, it would ensure the required homogenize mixing of the reactants that also avoid any of the mass transfer limitation. The circumferential configuration of the entry and exit of the reactor enables it to maintain a long residence time, even at a higher inlet mass flow that forces the flow to move in a swirling trajectory.

| Mesh type | Coarse | Baseline (Baseline × 1.5) | Fine mesh (Baseline × 1.89) |
|---|---|---|---|
| No. of cells | 281,382 | 427,742 | 534,228 |
| Residence time (sec) | $1.01 \times 10^3$ | $1.080 \times 10^3$ | $1.100 \times 10^3$ |
| Error % | 10% | 2% | — |

**Table 5.** Cells for the two meshes for the mesh sensitivity analysis.

### 5.3. Reactive flow analysis

The reactor tube is setup vertically and is subjected to inlet flow rate at Re = 6000 at the bottom of the tube into the inner reactor chamber. There, the two fluids of TG and methoxide (mixture of the methanol and the catalyst) are injected circumferentially at the stipulated molar/mass ratio and inlet temperature. The outflow is subjected to atmospheric pressure which is admitted at exit boundary condition of the outer tube chamber located at the top. The no-lip no penetration and insulated wall is also applied to all the bounded reactor walls.

To maintain single-phase flow avoiding the complication of other reactivation, the case of reactive flow simulation is carried out below the boiling point of methanol at a temperature of 60°C (333 K). A steady-state solution is sought for the flow. This is achieved by ignoring the temporal term of the governing equation. The flow is introduced to the reactor by means of an external peristaltic or diaphragm pump operating at relatively low head of nearly 2-m to overcome the viscous shear stresses and head losses and the reactor vertical head at an adjustable discharge capacity of up to 5 L/min.

### 5.4. Simulation results

Based on the stoichiometry of transesterification reactive 3 moles of methanol are consumed with 1 mole of triglycerides for the making of 1 mole of biodiesel. Using this ratio(3:1 methanol to WCO molar), which also corresponds to AL/WCO mass fraction of 0.102/0.892, results on the species contour plots depicted in **Figure 8** and the outlet fractions are summarized in **Table 6**.

At this velocity and molar ration, a low conversion of 28% is achieved. The conversion is described by Eq. (16) and is written as:

$$Conv = \frac{C_{0,WCO} - C_{f,WCO}}{C_{0,WCO}} \times 100 \tag{16}$$

where $C$ is the concentration and the subscripts $o$ and $f$ signify the initial and final state of the concentration of the waste cooking oil (WCO). When the model is subjected to higher molar ratio of 6:1 AL to WCO and at higher inlet mass flow results are depicted in **Figure 9** of all the participating species. This figure illustrates the reaction takes place only within the inner tube ring. The conversion seems to be enhanced reaching as high as 89% based on the inlet and exit fraction evaluated in the inner loop of the tubular cylinder, however a considerable intermediate is still beside TG conversion and mandating the need for the flow to duel and additional residence time to complete the reaction.

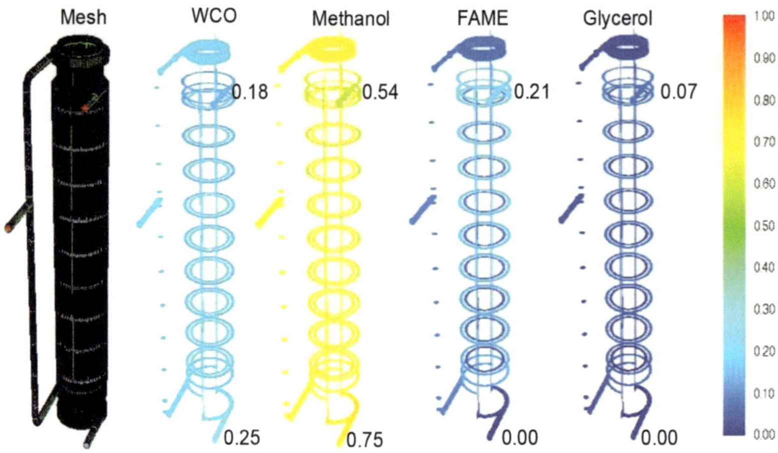


**Figure 8.** Molar fraction of species across the reactor (global scale).

| Port & ratio/species | $C_{54}H_{104}O_6$ (WCO) | $CH_4O$ | $C_{18}H_{36}O_2$ (FAME) | $C_3H_8O_3$ | Conversion |
|---|---|---|---|---|---|
| Inlet | 0.25 | 0.75 | 0 | 0 | 28% |
| Outlet | 0.18* | 0.54 | 0.21 | 0.07 | |

*Conversion = (0.25−0.18)/0.25.

**Table 6.** Species molar fraction at idealistic conditions.

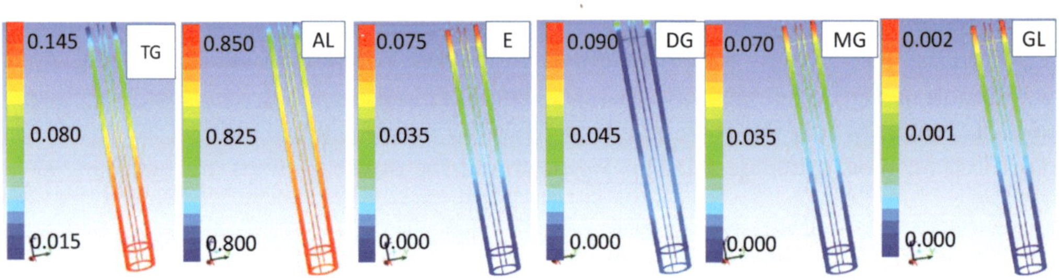


**Figure 9.** Molar fraction of the participating species within the first inner tube of the reactor and at 6:1 AL to WCO ratio.

From **Figure 9**, the concentration of the TG is reduced as the flow climbs up in the reactor as does the AL due to the production of biodiesel (E) and the two-intermediate species (MG and DG). The figure also clearly shows how DG precedes the formation of the MG and GL observed in the fading blue color contours of the DG concentration near the bottom inlet. The MG and GL show progressive color contours from the nil value represented by dark blue contours and reaching to higher value of 0.07 and 0.02 for each of MG and GL, respectively. Therefore, this allows additional residence time for the reaction in the second reactor tube reduces these intermediates and lead to higher production of the biodiesel and its byproduct

| Port & ratio/Species | $C_{54}H_{106}O_6$ (WCO) | $CH_4O$ | $C_{18}H_{36}O_2$ (FAME) | $C_3H_8O_3$ | Conversion |
|---|---|---|---|---|---|
| 3:1 methanol-WCO molar ratio | | | | | |
| Inlet | 0.25 | 0.75 | 0 | 0 | 0.66 |
| Outlet | 0.085 | 0.254 | 0.496 | 0.165 | |
| 6:1 methanol-WCO molar ratio | | | | | |
| Inlet | 0.143 | 0.857 | 0 | 0 | 0.66 |
| Outlet | 0.049 | 0.574 | 0.283 | 0.094 | |
| 9:1 methanol-WCO molar ratio | | | | | |
| Inlet | 0.10 | 0.90 | 0 | 0 | 0.66 |
| Outlet | 0.034 | 0.702 | 0.198 | 0.066 | |

**Table 7.** Species molar fraction at practical conditions.

glycerol. It must also be noted that these reactions are reversible and hence they can go either way, such that high conversion in the 1st tubular loop may be counter reacted by the second tubular loop and marking lower conversion.

Nevertheless, several sensitivity studies considering AL/WCO ratio, temperature, speed or mass flux of the feedstock etc. can be conducted. Results of the species concentration subjected to higher methanol ratio and larger influx of the feedstock of 600 mL/min (Re = 6000) are listed in **Table 7**. This result clearly marked a reasonable conversion of 66% which is higher than those obtained at lower influx %. The conversion seems also insensitive for the additional amount of the methanol which sounds counterintuitive initially and defies the experimental observation; this trend may be explained by the used kinetics which are inherited from the experimental work obtained at one ratio. These obtained values failed to capture the influence of the increase of the concentration of specific reactant to steer the reaction forward. Therefore, this may suggest another testing procedure at another molar ratio in parallel, mimicking those obtained and shown in **Figure 4**. Though, results emphasize the advantage of increasing residence time and efficient mixing of the flow which reduces the additional methanol concentration.

## 6. Conclusions

Transesterification of WCO to biodiesel is presented in this work and experimental and numerical analysis to this green process is presented. Initially, the chemical kinetics for the reaction are evaluated to the multiple reactions of the transesterification and in both the forward and reversal pathways for these reactions. The obtained kinetics presented by the activation energy (E) and the reaction pre-constant are evaluated and used to establish a high fidelity and robust reactive flow model. This model is based on computational fluid dynamics that is governed by the Navies-Stokes equations for isothermal, multiple species reactive flow in turbulent regime. The model is built around the newly patented multiple tube and continuous transesterification reactor. The flow enters the reactor laterally and induces swirling flow. In turn, this results in an order of magnitude residence time higher than the transfer time that is based on the tube length and inlet velocity. Reactive flow at low velocity results in low conversion of the WCO, while actual turbulent flow significantly increases the conversion rate. Excess methanol mass sensitivity was insignificant to the conversion parameter that suggest that modeling can be still limited and suggesting to integrate more kinetics data. This work emphasizes the efficient mixing of the flow, interplaying as a parameter to the additional methanol concentration, thereby avoiding their downstream separation. In closing, numerical simulation of transesterification undoubtedly demonstrates the effectiveness of this tool in analyzing complicated reactive flow as the student faces in transesterification. Still, the process fails to reach completion even at higher reactant concentration and beyond the process stoichiometry. This creates an opportunity in this area to go beyond the conventional methods, such as electrical stimulation or signification and to assess their needed kinetics accordingly. The results demonstrate the feasibility of reactive flow dynamics in capturing and numerically simulating the transesterification process. This work substantiates the practicability of using numerical methods to construct precise and insightful image of the distribution of reaction rates and associated species to design and develop more efficient reactors.

## Acknowledgements

Khalifa University of Science and Technology, Masdar Institute Campus is highly acknowledged for their support.

## Author details

Isam Janajreh[1]* and Manar Almazrouei[2]

*Address all correspondence to: isam.janajreh@ku.ac.ae

1 Khalifa University of Science and Technology, Masdar City, Abu Dhabi, United Arab Emirates

2 Mechanical Engineering - (COE), United Arab Emirates University, Al Ain, UAE

## References

[1] Pradhan A, Shrestha DS. Impact of some common impurities on biodiesel cloud point. In: 2007 ASAE Annual Meeting. American Society of Agricultural and Biological Engineers; 2007. p. 1

[2] Banga S, Varshney PK. Effect of impurities on performance of biodiesel: A review. Journal of Scientific and Industrial Research. 2010;**69**(8):575-579

[3] Li Z, Ramanathan S, Chen C-C. Predicting thermophysical properties of mono-and diglycerides with the chemical constituent fragment approach. Industrial & Engineering Chemistry Research. 2010;**49**(11):5479-5484

[4] Ceriani R, Goncalves CB, Coutinho JAP. Prediction of viscosities of fatty compounds and biodiesel by group contribution. Energy & Fuels. 2011;**25**(8):3712-3717

[5] Noureddini H, Zhu D. Kinetics of transesterification of soybean oil. Journal of the American Oil Chemists' Society. 1997;**74**(11):1457-1463

[6] Janajreh I, ElSamad T, AlJaberi A, Diouri M. Transesterification of waste cooking oil: Kinetic study and reactive flow analysis. Energy Procedia. 2015;**75**:547-553

[7] Janajreh I, Al Shrah M. Numerical simulation of multiple step transesterification of waste oil in tubular reactor. Journal of Infrastructure Systems. 2016;**22**(4):A4014007

[8] Janajreh I, Almusharekh M, Ghenai C. Transesterification process of waste cooking oil: Catalyst synthesis, kinetic study, and modeling sensitivity. In: International Conference on Sustainable Solid Waste Management. 2014

[9] Janajreh I, ElSamad T. Mohammed Noorul Hussain, intensification of transesterification via sonication numerical simulation and sensitivity study. Applied Energy. 2017;**185**:2151-2159

[10] Janajreh I, Al Musharrekh M, Hussain M. Numerical simulation of transesterification and sensitivity study. International Journal of Modern Engineering. 2015;**15**(2):13-22

[11] Laidler KJ. The development of the Arrhenius equation. Journal of Chemical Education. 1984;**61**(6):494

[12] Abd Rabu R, Janajreh I, Honnery D. Transesterification of waste cooking oil: Process optimization and conversion rate evaluation. Energy Conversion and Management. 2013;**65**:764-769

[13] MATLAB. version 8.3.0 (R2014a). Natick, Massachusetts: The Math Works Inc.; 2014

[14] Freedman B, Butterfield RO, Pryde EH. Transeterfication kinetics of soybean oil. Journal of the American Oil Chemists' Society. 1986;**63**:1375-1380

[15] Freedman B, Pryde EH, Mounts TL. Variables affecting the yield of fatty esters from transesterfified vegetable oils. IBID. 1984;**61**:1638-1643

[16] Farkas L, Schachter O, Vromen BH. On the rate of acetate. Journal of the American Chemical Society. 1949;**71**:1991-1994

[17] Dufeck EJ, Butterfield RO, Frankel EN. Esterification and transsesterification of 9(10)-caboxystearic acid and its methyl esters, kinetic studies. Journal of the American Oil Chemists' Society. 1972;**49**:302-306

[18] Ansys Fluent 17.1. ANSYS Inc. Products; 2013

[19] Janajreh IM, Rabu RA. System and method for continuous transesterification of oils, US Patent 9,468,899; 2016

[20] Laidler K. The development of the Arrhenius equation. Journal of Chemical Education. 1984;**61**(6):494

[21] Narvaez PC, Rincon SM, Castaneda LZ, Sanchez FJ. Determination of some physical and transport properties of palm oil and of its methyl esters. Latin American Applied Research. 2008;**38**:1-6

[22] Cokljat D, Slack M, Vasquez SA. Reynolds-stress model for Eulerian multiphase. In: Hanjalic YNK, Tummers MJ, editors. Proceedings of the 4th International Symposium on Turbulence Heat and Mass Transfer. Begell House, Inc.; 2003. pp. 1047-1054

[23] Hinze JO. Turbulence. New York: McGraw-Hill Publishing Co.; 1975

[24] Fox RO. Computational Models for Turbulent Reacting Flows. Cambridge, England: Cambridge University Press; 2003

Section 2

# Applications and Challenges

Chapter 5

# Measurement of Limited and Unlimited Emissions during Burning of Alternative Fuels in the Tractor's Engines

Juraj Jablonický, Danela Uhrínová, Juraj Tulík and Ján Polerecký

Additional information is available at the end of the chapter

http://dx.doi.org/10.5772/intechopen.79705

**Abstract**

This text is aimed at the basic analysis of diesel oil and rapeseed methyl ester and evaluation of limited and unlimited emission produced by their combustion. Thereafter, test results are compared, and the evaluation of emission—greenhouse gases, dangerous exhaust gases and strong carcinogens and their contents during fuel combustion—is done. In this chapter, results obtained from the application of biofuel to the machinery working in conditions sensitive to environmental contamination are presented. At present, our environment is excessively overloaded with all kinds of emission, and the idea of using fuel with a marginal impact on the environment is very important. Based on the evaluation of emission, it can be stated that it is very important to study not only limited but also unlimited emission that can be very dangerous, although in this work it was discovered that values of unlimited emission do not exceed the lethal limit.

**Keywords:** alternative fuels, exhaust, internal combustion engine, tractor, limited emissions, unlimited emissions

## 1. Introduction

Developments in the field of technology also bring with their positive impact and increased quality of life undesirable side effects. One of the severe adverse effects of scientific and technical progress is environmental pollution [1–3]. We can now witness the producers' efforts to increase the ecological safety taking place as early as in the combustion engine development for motor vehicles. As a result, vehicles are equipped with exhaust systems supplemented with exhaust gas treatment equipment. Engine developers and producers can also utilize alternative fuels (CNG, LPG, E85, biodiesel) as an energy source for vehicles, as well as transition to hybrid propulsion (combustion and electric motor) and introduction of fuel cells. The authors Vitázek et al. deal with gas emissions resulting from the combustion of biofuels in the environment [4, 5].

With regard to the need for limiting air pollution by combustion engine exhaust gases, the maximum production of pollutants in exhaust gas emissions of all motor vehicles is legally limited. From a legislative point of view are limited emissions of CO, hydrocarbons, volatile organic HC compounds, suspended PM, and NOx oxides. Engine emissions contain hundreds of chemicals with different concentrations, the biological properties (impact on human health) of which have not been exactly determined to date. Combustion engines are responsible for more than 70% of global CO emissions production and 19% of global $CO_2$ emissions production [6]. In addition to the products of perfect combustion—i.e., $CO_2$, $H_2O$, excess oxygen, excess nitrogen—which represent the majority of exhaust products, a wide range of other gases and solids may occur, which tend to receive greater attention: CO, non-combusted hydrocarbons (paraffins, olefins, aromatic hydrocarbons), partially combusted hydrocarbon (aldehydes and ketones), degradation products (acetylene, ethylene, hydrogen, soot), nitrogen oxides NOx (NO—nitrogen monoxide, $N_2O$—dinitrogen monoxide, $NO_2$—nitrogen dioxide) and solid particulate matter. It is evident from the operation of a compression ignition engine that the increase in smoke opacity leads to an increase in production of pollutants (CO and HC), the measuring of which is difficult in practice, however. It is therefore vital to observe the value of particulate emissions (PM, particulate matter), the measuring of which is much faster and technologically and economically simple, while simultaneously being sufficient for evaluation of technical condition. Compression ignition engines operate narrowly below the smoke opacity threshold during the maximum performance [7].

One of the ways to comply with the stricter emission regulations is to focus on and search for suitable alternative fuels, as suggested by Ulusoy et al. [8], stating that the main plausible alternative fuels used in car transport are ethanol, hydrogen and biodiesel. A large number of studies have shown that biodiesel could serve as an alternative for compression ignition engines, with small or even no requirements for their adjusting [9]. It was also proven that biodiesel has a great potential for decreasing the $CO_2$, CO, THC (total hydro carbon) emissions and PM (particulate matter) emissions [10, 11]. In conclusion, alternative fuels and their mixtures with diesel fuel are still a subject of research, focusing primarily on reduction of emission arising from their combustion in engine, and also on the transformation of their heat into mechanical energy.

Agriculture is part of nature and the countryside. Ecological agriculture and environmental protection are the world's global interests [12, 13]. There are a lot of negatives on fossil fuels, on which our society is depending to a high degree. One of the most important disadvantages is fouling the air and causing greenhouse effect, which affects weather with regard to temperature [14, 15]. This work deals with a partial alternate use of diesel oil from a renewable fuel—rapeseed methyl ester (RME). An analysis based on limited and unlimited emission detection was performed.

## 2. Materials and methods

The measuring standard was an international standard used for non-road engines. According to International Organization for Standardization (ISO), this standard specifies the test cycles for the measurement and evaluation of gaseous and particulate exhaust emission from reciprocating internal combustion engines, and it is applicable to engines for mobile, transportable and stationary use [16–18]. Characteristics of 8-point cycle by ISO 8178–4, C1 are shown in **Figure 1** [20].

The conversion of individual compounds of exhaust gases from ppm to g.kW$^{-1}$ h$^{-1}$:

$$ZL_i = EVP_{is} \cdot \frac{Mm_i . t_{vps}}{Mm_{vps} . P_e} = EVP_{iv} \cdot \frac{Mm_i . t_{vpv}}{Mm_{vpv} . P_e}, g.kW^{-1} h^{-1} \tag{1}$$

*clarity:* $ZL_i$—concentration of contaminant linked with effective power, g.kW$^{-1}$ h$^{-1}$, $EVP_{is,id}$—emission of exhaust gases (dry—s, moist—v) of compound i, as volume unit share, ppm; $M_{mi}$—minor mass of compound i, kg.kmol$^{-1}$; $Mm_{vps}$—minor mass of exhaust gases (dry), kg.kmol$^{-1}$; $Mm_{vpv}$—minor mass of exhaust gases (moist), kg.kmol$^{-1}$; $t_{vps}$—mass flow of exhaust gases (dry) kg.h$^{-1}$; $t_{vpv}$—mass flow of exhaust gases (moist), kg.h$^{-1}$. For the first, second and third points of measurements—0.15; for the fourth, fifth, sixth and seventh points—0.1; for the eighth point—0.15; $P_e$—effective power, kW.

Needed data for conversion from ppm to g.kW$^{-1}$ h$^{-1}$ are listed in **Table 1** [20].

### 2.1. Measured objects and measured devices

The technical description of both objects measured is specified in **Table 2** and of measured devices is shown in **Table 3** [20].

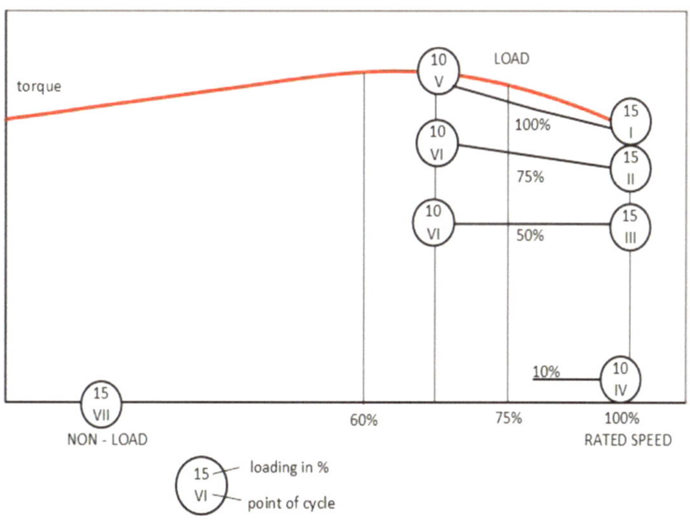


**Figure 1.** Characteristics of eight-point cycle by ISO 8178–4, C1.

| Substance, *i* | | Mass ($kg.kmol^{-1}$) | Note |
|---|---|---|---|
| $M_{mi}$ | $NO_2$ | 46.0060 | NOx process as $NO_2$ |
| | CO | 28.0104 | |
| | HC | 13.8760 | HC 1 |
| | $SO_2$ | 64.0610 | |
| $Mm_{vps}$ | Exhaust gases—dry | 30.21/29.84 | 5% $O_2$/9.6 $O_2$ |
| $Mm_{vpv}$ | Exhaust gases—moist | 28.84/28.82 | 5% $O_2$/9.6 $O_2$ |

**Table 1.** Needed data for conversion from ppm to $g.kW^{-1}\ h^{-1}$.

| Tractor 1 (turbocharged) | | Tractor 2 | |
|---|---|---|---|
| Engine | S. L. H—H 100.4 WT | Engine | Deutz 2012, TCD 2012 L04 2 V |
| Number of cylinders | 4 | Number of cylinders | 4 |
| Capacity of cylinders | 4.00 $dm^3$ | Capacity of cylinders | 4.038 $dm^3$ |
| Cavil/stroke | 105 mm/115.5 mm | Cavil/stroke | 101 mm/126 mm |
| Rated revolution | 2500 $min^{-1}$ | Rated revolution | 2300 $min^{-1}$ |
| Power | 65 kW | Power | 72.5 kW |
| Emission class | Stage I | Emission class | Stage III A |

**Table 2.** Technical data of the tractors.

| Device name | Device producer |
|---|---|
| Dynamometer | Schenck W700 |
| Emission testing system | AVL—SESAM 4 (FTIR) |
| Fuel consumption | AVL 733S |

**Table 3.** Used measured devices.

## 3. Results

### 3.1. Measurement of technical properties of tractors

Measurement was carried out with the use of two fuels. First type was the diesel fuel, which was used as a reference sample for diesel RME comparisons. Measurements results were evaluated and compared reciprocally. Used measuring devices were connected according to the scheme shown in **Figure 2**. Processed measurement results are shown in **Figure 3**. We can see the appliance connection together with measured parameters. The measurements were repeated three times for both tractors, with each fuel, at full load in accordance to OECD

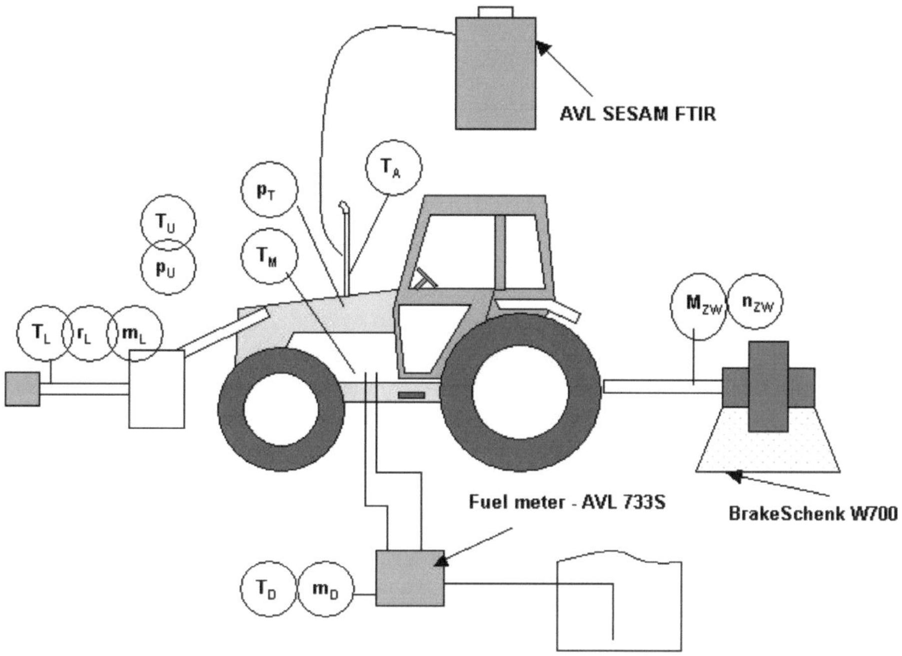


**Figure 2.** View of test equipment (Müllerová, Landis, Schiess: Agroscope Reckenholz-Tänikon Research Station ART and SUA in Nitra).

Code2 test. Measure points were set at 1000 $min^{-1}$ of PTO revolutions. Based on results curve at full load, the calculation of measure points for emission measurement was processed. These measure points comply with the ISO 8178–4 standard. According to this standard, each measure point lasts for 10 min. Emission test system AVL-SESAMFTIR uses two methods of calculation—diesel and biodiesel. These are set according to the fuel used. The value of smokiness is evaluated at 95% and 70% nominal evolutions and at maximal torque.

### 3.2. Measurements of limited emission

Measurement of CO, HC, NOx and particulate emissions on both tractors were done according to ISO 8178-4, $C_1$ - 8 points (**Figures 4** and **5**), [20]. The conversion from *ppm* to $g.kW^{-1}\,h^{-1}$ was made by pattern (1) using the values from **Table 1**. In **Table 4**, there are figured standard deviations from three repetitions. The graphic description of limited emission is represented in **Figures 6–11**.

Based on measured values of limited emission, average value by the next pattern was calculated:

- the average value of sign in a subgroup [19]:

$$\overline{x}_i = \frac{1}{n} \tag{2}$$

*clarity*: *i* = *1,2,....,* *k* and *j* = *1,2,....n.*

- the standard deviation in a subgroup [19]:

$$s_i = \sqrt{\frac{1}{n-1}\sum_{j=1}^{n} x_{ij}} \quad (3)$$

*clarity*: $i = 1,2,...,k$ and $j = 1,2,..., n$; $i$—marking of a subgroup; $j$—numerical order of measured value in a subgroup, $n$—range of subgroup, $X_{ij}$—measured value in a $i$–subgroup.

- the average value X:

$$\overline{\overline{x}} = \frac{1}{k} \quad (4)$$

- the average of standard deviations of individual subgroups [19]:

$$\overline{s} = \frac{1}{k} \quad (5)$$

Based on the patterns above, the standard deviations of limited emission measured on Tractor 1 with turbo and Tractor 2 were determined. The standard deviations of individual emissions

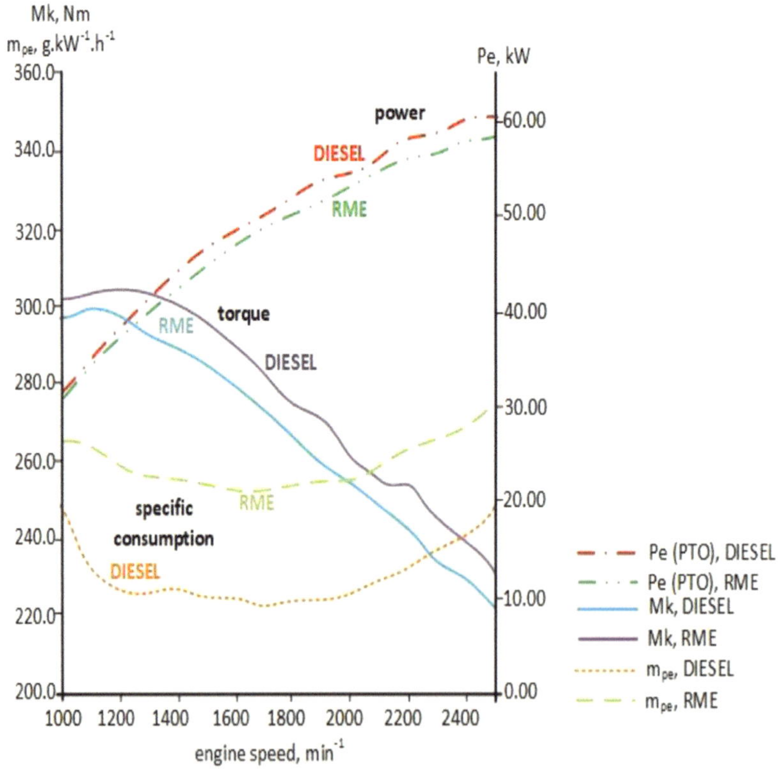


**Figure 3.** Engine-speed map of Tractor 1 (turbocharged) with fuels—diesel and biofuel RME (Müllerová, Landis, Schiess: Agroscope Reckenholz-Tänikon Research Station ART and SUA in Nitra).

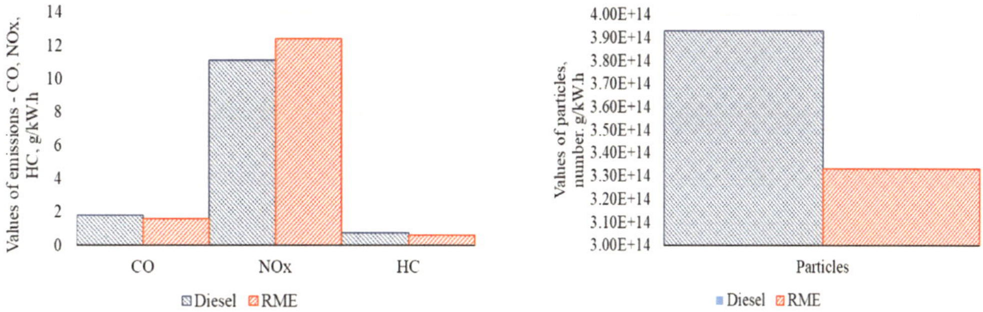


**Figure 4.** Values of limited emission for Tractor 1 (turbocharged).

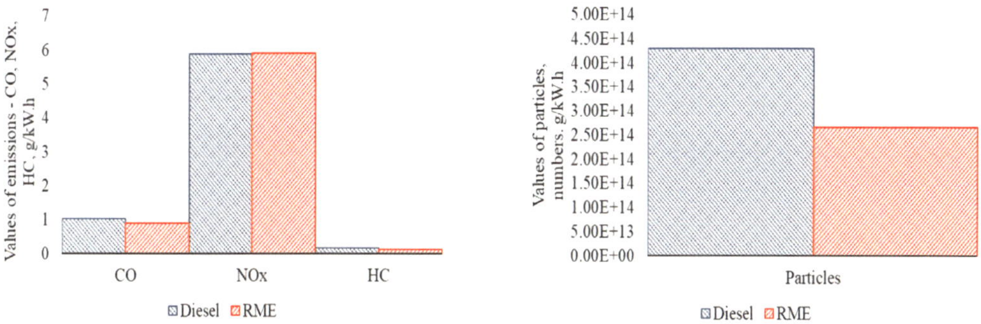


**Figure 5.** Values of limited emission for Tractor 2.

| | CO | NOx | HC | Particles |
|---|---|---|---|---|
| | ($g.kW^{-1}.h^{-1}$) | | | (Number. $kW^{-1}.h^{-1}$) |
| **Tractor 1 (turbocharged)** | | | | |
| Diesel | 1.80 | 11.13 | 0.77 | 3.93E + 14 |
| RME | 1.61 | 12.42 | 0.60 | 3.33E + 14 |
| **Tractor 2** | | | | |
| Diesel | 1.05 | 5.90 | 0.19 | 4.31E + 14 |
| RME | 0.91 | 5.92 | 0.13 | 2.66 + E14 |

*Average value, based on PTO power.

**Table 4.** Values of limited emission*.

are demonstrated in **Figures 6–11**. In **Figures 6–11**, reciprocally compared standard deviations of limited emission measured using diesel oil and RME are given [20].

The values of CO and HC and also particle emission are lower for RME as in **Table 4**, but the values of $NO_x$ are lower for diesel oil. It is evident that the newer engine of Tractor 1 decreases

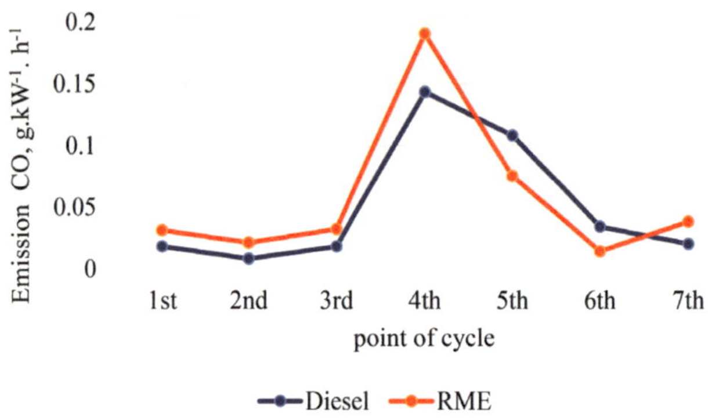


**Figure 6.** Standard deviation values of limited emission—CO for Tractor 1 (turbocharged) (Müllerová, Landis, Schiess,: Agroscope Reckenholz-Tänikon Research Station ART and SUA in Nitra).

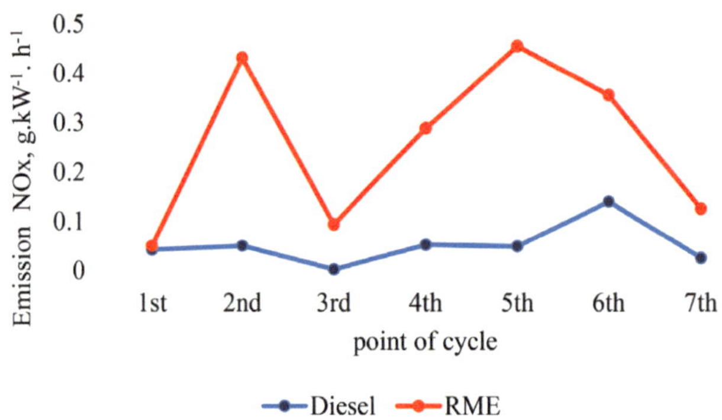


**Figure 7.** Standard deviation values of limited emission—NOx for Tractor 1 (turbocharged) (Müllerová, Landis, Schiess: Agroscope Reckenholz-Tänikon Research Station ART and SUA in Nitra).

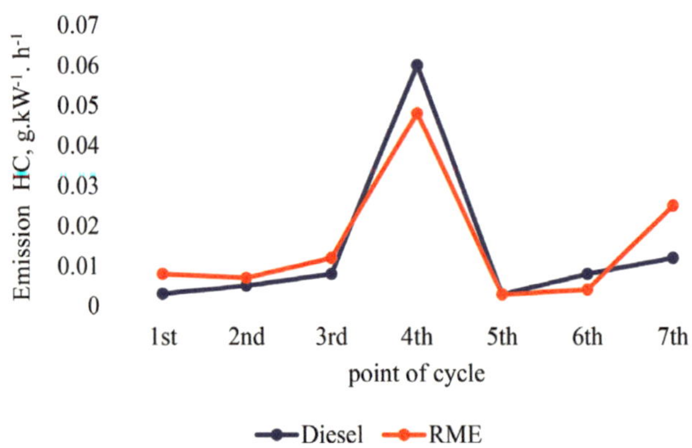


**Figure 8.** Standard deviation values of limited emission—HC for Tractor 1 (turbocharged) (Müllerová, Landis, Schiess: Agroscope Reckenholz-Tänikon Research Station ART and SUA in Nitra).

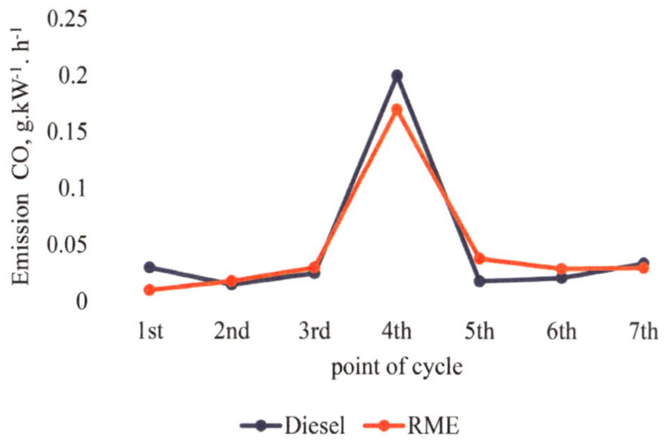


**Figure 9.** Standard deviation values of limited emission—CO for Tractor 2 (Müllerová, Landis, Schiess: Agroscope Reckenholz-Tänikon Research Station ART and SUA in Nitra).

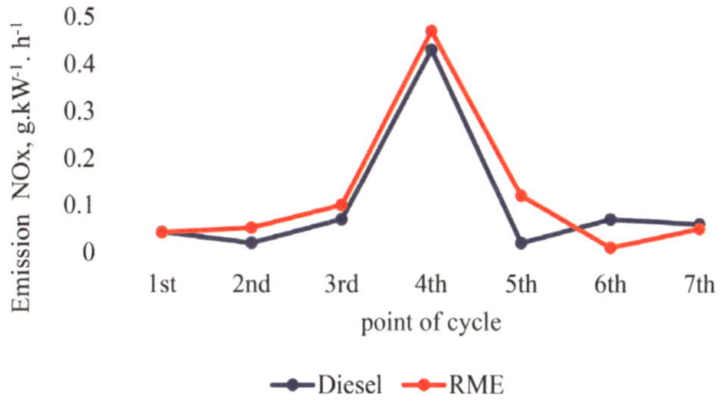


**Figure 10.** Standard deviation values of limited emission—NOx for Tractor 2 (Müllerová, Landis, Schiess: Agroscope Reckenholz-Tänikon Research Station ART and SUA in Nitra).

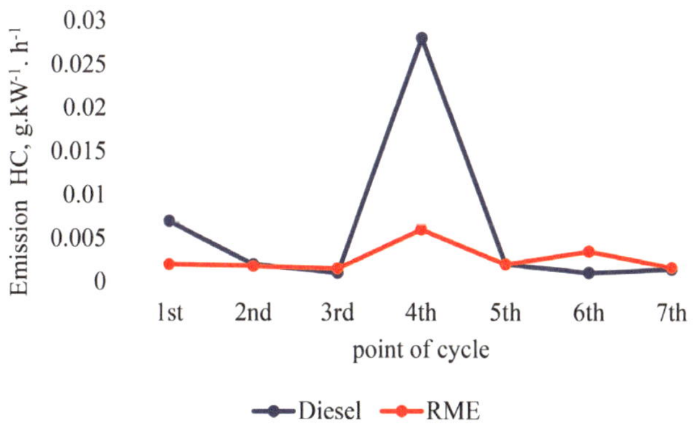


**Figure 11.** Standard deviation values of limited emission—HC for Tractor 2 (Müllerová, Landis, Schiess: Agroscope Reckenholz-Tänikon Research Station ART and SUA in Nitra).

| Tractor 1 turbocharged | CO2 | NO | NO2 | N2O | NH3 | CH4 |
|---|---|---|---|---|---|---|
| ppm | Carbon dioxide | Nitric oxide | Nitrogen dioxide | Nitrous oxide | Ammonia | Methane |
| Diesel | 55,867 | 845 | 40 | 0.5 | 0.13 | 0.52 |
| RME | 56,769 | 890 | 43 | 0.66 | 0.21 | 1.27 |
| | C4H6 | HCN | AHC | SO2 | HCHO | MECHO |
| | 1,3 Butadiene | Hydrogen cyanide | Aromatic HC | Sulfur dioxide | Formaldehyde | Acetaldehyde |
| Diesel | 0.97 | 0.57 | 2.1 | 4.6 | 8.1 | 2.7 |
| RME | 1.98 | 0.57 | 1.19 | 1.40 | 9.95 | 0.57 |
| Tractor 2 | CO2 | NO | NO2 | N2O | NH3 | CH4 |
| ppm | Carbon dioxide | Nitric oxide | Nitrogen dioxide | Nitrous oxide | Ammonia | Methane |
| Diesel | 64,426 | 378 | 16.9 | 0.43 | 0.12 | 0.1 |
| RME | 66,040 | 431 | 16.8 | 0.57 | 0.13 | 0.1 |
| | C4H6 | HCN | AHC | SO2 | HCHO | MECHO |
| | 1,3 Butadiene | Hydrogen cyanide | Aromatic HC | Sulfur dioxide | Formaldehyde | Acetaldehyde |
| Diesel | 0.44 | 0.59 | 0.77 | 5.0 | 2.23 | 0.49 |
| RME | 0.90 | 0.45 | 1.19 | 2.9 | 2.08 | 0.73 |

**Table 5.** Values of unlimited emission (Müllerová, Landis, Schiess,: Agroscope Reckenholz-Tänikon Research Station ART and SUA in Nitra).

emission significantly. Measured values are based on PTO power, so cannot be evaluated by emission standards for off-road vehicles. If these measurements were done on engine, both tractors will meet the emission norm for CO and HC of RME and diesel. The values of $NO_x$ are higher about 21% for both fuels tested in agricultural Tractor 1 (turbocharged) and about 25% for fuels tested in agricultural Tractor 2.

### 3.3. Measurements of unlimited emission

Measurements of unlimited emission were also done, which are possible to measure by AVL SESAM FTIR 4—$CO_2$, NO, $NO_2$, $N_2O$, $NH_3$, $CH_4$, $C_4H_6$, HCN, AHC, $SO_2$, HCHO and MECHO. In **Table 3**, average values from three repetitions for each fuel (diesel oil, RME) are figured. The tractor that used RME had not only higher values of $NO_x$ (NO, $NO_2$ and $N_2O$) but also almost 50% higher values of ammonia, methane and 1.3-butadiene, which are considered to be dangerous substances. The newer engine of Tractor 2 had higher values of $NO_x$, acetaldehyde and 1.3-butadiene for RME but the difference was not so big.

On the other side, lower values with RME for sulfur dioxide and acetaldehyde were obtained for Tractor 1 (turbocharged) and for sulfur dioxide, hydrogen cyanide and formaldehyde for

http://dx.doi.org/10.5772/intechopen.79705

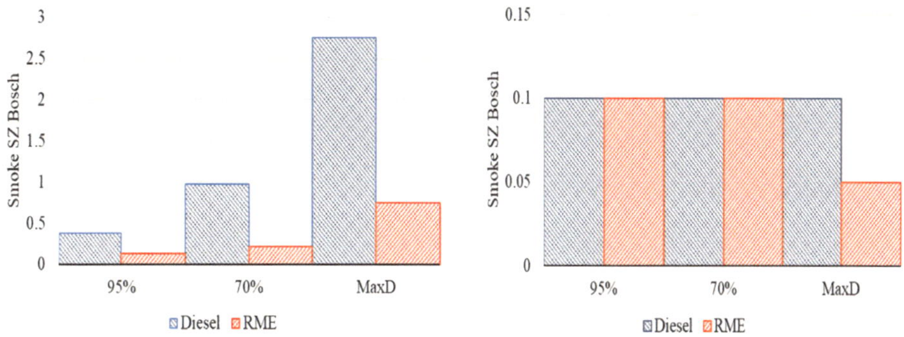


**Figure 12.** Measurement of smoke (Müllerová, Landis, Schiess: Agroscope Reckenholz-Tänikon Research Station ART and SPU Nitra).

Tractor 2. Nevertheless, the values of unlimited emission are negligible, except carbon dioxide whose values were higher with RME for both tractors.

Values of unlimited emission are shown in **Table 5**.

### 3.4. Measurements of emitted smoke

The values of smoke in exhaust gases are usually a lot lower with RME than diesel oil. For Tractor 1 (turbocharged), the value of smoke was more than 50% lower with RME than diesel oil. From **Figure 12**, it is evident that Tractor 2 had much lower value of smoke. These values were near zero and it did not matter whether RME or diesel oil was used.

## 4. Conclusion

At present, when our environment is overburdened with emissions of all kinds, the idea of using fuels with minimal impact on the environment is of great importance. Fuel produced from methyl ester of vegetable oils can be considered advantageous in that almost every diesel engine is in principle capable of combusting such fuels. Taking into account the fact that up to 90% of the transport of passenger and cargo transport is provided by diesel fuel vehicles (trucks, buses, locomotives, ships, tractors) at the present, fuels made from methyl ester of vegetable oils represent a huge potential for their use in the diesel combustion engines. Very often, the emphasis is on the contribution of fuels of plant origin in terms of creating a carbon balance in nature. The production of carbon dioxide during combustion corresponds to its consumption during photosynthesis. Biodegradability, for example, rapeseed oil methyl ester after its release into the environment, is approximately 95%.

In this chapter, the results reached from application of biofuel to the machinery working in condition that are sensitive to environment contamination are presented. At present, our environment is excessively overloaded with all kinds of emission and the idea of using fuel with marginal impact on environment is very important. It is possible to state that the differences of these two tractors are peculiar to their engines' construction, the year of production and specification (Tractor 2 is specified as 100% biodiesel). By evaluation of emission (GHG, dangerous exhaust gases and carcinogens) it can be declared that it is very important to study not just limited but also unlimited emissions, which can be very dangerous, although in this work it was discovered that values of unlimited emission do not exceed lethal limits.

## Acknowledgements

This work was supported by Ministry of Education, Science, Research and Sport of the Slovak Republic, Project VEGA 1/0464/17 "Monitoring of the impact of ecological fuels obtained from the agricultural production and additives in hydrocarbon fuels to technical and environmental performance of internal combustion engines used in agricultural and transport technique."

This work was supported by AgroBioTech Research Centre built in accordance with the project Building "AgroBioTech" Research Centre ITMS 26220220180.

## Author details

Juraj Jablonický[1]*, Danela Uhrínová[1], Juraj Tulík[1] and Ján Polerecký[2]

*Address all correspondence to: juraj.jablonicky@gmail.com

1 Department of Transport and Handling, Faculty of Engineering, Slovak University of Agriculture in Nitra, Slovak Republic

2 S-EKA Ltd. is a Designated Service (Technical Service) for the Emission Control of Road Motor Vehicles, Nitra, Slovak Republic

## References

[1] Jablonický J, Tkáč Z, Majdan R, Uhrinová D, Hujo Ľ, Vozárová V. Properties Evaluation of Biofuels and Bio-lubricants. Nitra: Slovak University of Agriculture; 2012

[2] Králik M, Jablonický J, Nikolov MI. Monitoring of NOx Emissions at Selected Diesel Engine. Angel Kanchev University of Ruse; 2015

[3] Lenďák P, Jablonický J, Uhrinová D, Kosiba J, Polerecký J. Possible solutions for checking particulate matter filters (DPF) in motor vehicles. Advanced Materials Research. 2014;**1059**(spec. iss):119-125

[4] Vitázek I et al. Gas emission during combustion biofuel. Acta Technologica Agriculturae. 2014;**17**(3):75-79. ISSN 1335-2555

[5] Vitázek I et al. Production of emission during combustion of various biofuels. Agricultural Engineering. 2014;**4**:41-50. ISSN 0554-5587

[6] Tauzia X, Maiboom A, Shah SR. Experimental study of inlet manifold water injection on combustion and emissions of an automotive direct injection diesel engine. Energy. 2010;**35**(9):3628-3639

[7] Vlk F. Automotive Combustion Engines. Brno: Nakladatelství a vydavatelství VLK; 2003. ISBN: 80-238-8756-4

[8] Ulusoy Y, Arslan R, Kaplan C. Emission characteristics of sunflower oil methyl ester. Energy Sources, Part A: Recovery, Utilization, and Environmental Effects. 2009;**31**(11): 906-910

[9] Hammond GP, Kallu S, McManus MC. Development of biofuels for the UK automotive market. Applied Energy. 2008;**85**(6):506-515

[10] Regulation no. 24 of the Economic Commission for Europe of the United Nations (UN/ECE). Uniform provisions concerning: I. The approval of compression ignition (C.I.) engines with regard to the emission of visible pollutants; II. The approval of motor vehicles with regard to the installation of C.I. engines of an approved type; III. The approval of motor vehicles equipped with C.I. engines with regard to the emission of visible pollutants by the engine; IV. The measurement of power of C.I. engines. 2018. http://eur-lex.europa.eu/legal-content/en/TXT/?uri=CELEX%3A42008X0222%2801%29

[11] Dorado MP, Ballesteros E, Arnal JM, Gómez J, López FJ. Exhaust emissions from a diesel engine fueled with transesterified waste olive oil. Fuel. 2003;**82**(11):1311-1315

[12] Cvíčela P, Drabant Š, Majdan R. Sledovanie prevádzkového zaťaženia hydrogenerátora UD 25 (Observation of operating loading of the hydrostatic pump UD 25). In: International Student Scientific Conference. Nitra: Slovak University of Agriculture in Nitra; 2008

[13] Gábriš Ľ. Ochrana a tvorba životného prostredia v poľnohospodárstve [Protection and Creation of Environment in Agriculture]. Nitra: Slovak University of Agriculture in Nitra; 1998

[14] Majdan R, Cvíčela P, Bohát M, Ivanišová K. The observation of hydrostatic pump deterioration during the durability test according to hydraulic fluids contamination. In: X. International Conference of Young Scientists 2008: Conference Proceedings, Czech Republic. Prague: Czech University of Life Sciences Prague; 2008. pp. 147-153

[15] Semetko J, Janoško I, Pernis P. Increase of driving pulls of the tractor by trailer. In: Actual Tasks on Agricultural Engineering: Proceedings of the 31st International Symposium on Agricultural Engineering, Opatija Croatia. Zagreb: Sveučilište u Zagrebu; 2003. pp. 59-64

[16] Tkáč Z et al. Alternatívne palivá pre motory [Alternative Fuels for Engines]. Nitra: Slovak University of Agriculture in Nitra; 2008

[17] Vlk F. Alternativní pohony motorových vozidel [Alternative Drives of Motor Vehicles]. Prof. Ing. Frantisek Vlk, DrSc., nakladetlství a vydavatelství. Brno; 2004. ISBN: 80-239-1602-5

[18] Šimor R, Janoško I, Čery M. Možnosti merania emisií spaľovacieho motora [Possibilities of Emission Measurement for Combustion Engine]. In: Technics in Agrisector Technologies. Nitra: Slovak University of Agriculture; 2008. p. 94104

[19] Hrubec J. Riadenie kvality [Quality Control]. Nitra: Slovak University of Agriculture in Nitra; 2001

[20] Uhrinová D, Jablonický J, Hujo Ľ, Kosiba J, Tkáč Z, Králik M, Chrastina J. Research of limited and unlimited emission effect on the environment during the burning of alternative fuels in agricultural tractors. Journal of Central European Agriculture. 2013, 2013;**14**(4):1402-1414. Online. Dostupné na internete: <http://jcea.agr.hr/articles/774439_RESEARCH_OF_LBestregardsIMITED_AND_UNLIMITED_EMISSION_EFFECT_ON_THE_ENVIRONMENT_DURING_THE_BURNING_OF_ALTERNATIVE_FUBest regardsELS_IN_AGRICULTURAL_TR_en.pdf>. ISSN: 1332-9049

Chapter 6

# Development and Implementation of Virtual Instrumentation for the Measurement of Operating Parameters of an Engine Using Diesel-Biodiesel Mixtures

Armando Pérez, Rogelio Ramos, Gisela Montero, Conrado García, Marcos Coronado, Héctor Campbell, René Delgado and Alejandro Suástegui

Additional information is available at the end of the chapter

http://dx.doi.org/10.5772/intechopen.80533

**Abstract**

Through a reaction of alkaline transesterification of soybean oil using sodium methoxide, biodiesel denominated as B100 was obtained, with which four mixtures of diesel-biodiesel B2, B5, B10, and B20 were prepared. Kinematic viscosity and high heating value of the four blends, B100, and diesel were determined. The blends, B100, and diesel were used in a motor of four cylinders in-line engine, air intake at atmospheric pressure with a power of 250 hp and 6000 cm$^3$, operating at a constant rate of 850 ± 50 rpm, a temperature of 25°C, and a relative humidity of 50%. To monitor the emissions, rpm, fuel consumption, and temperature in the engine's exhaust manifold, which operates with diesel-biodiesel mixtures, an integral instrument that uses the virtual instrumentation technology was developed in the programming platforms LabVIEW 2010 and ARDUINO. The development and implementation of the virtual instrument allow monitoring in real time the parameters of internal combustion engines and presents the versatility, flexibility, scalability, and capacity to function in equipment that operates with different liquid fuels at a lower cost than the one that conventional systems offered. These characteristics represent a significant benefit in comparison with the measurement and monitoring systems in the present market.

**Keywords:** biodiesel, virtual instrumentation, operations parameters, compression ignition engine

## 1. Introduction

The trend in fossil fuel consumption is increasing; adding to that, the emissions of NO, CO, $SO_2$, and $CO_2$ caused by the use of fuel oils like diesel affect directly and indirectly the environment, as well as the population quality of life [1–4]. Currently, engines with a highly developed torque are used for heavy work like urban transport, cargo transport, light passenger transport, agricultural machinery, emergency systems, craft and power generation powered by diesel, since diesel has a high heating value [5].

Some studies prove that increasing the percentage of biodiesel, that substitutes diesel, decreases the emissions produced by its combustion [6, 7]. The biodiesel is considered a renewable and ecological fuel drawn from lipids that are made to react with short-chain alcohol in the presence of a catalyst that can be acid, base or enzymatic, to produce a fatty acids' mono-alkyl esters mixture [8].

The biodiesel contains oxygen in its molecule, which helps to increase the combustion efficiency inside the compression ignition engine's (CIE) combustion chamber [9]. The European Union is the leading producer of biodiesel across the world. In the case of America are the United States, Argentina and Brazil, who use soy, corn, canola, rapeseed, and palm oil, tallow yellow and white fat as raw materials.

In 2013, the production of biodiesel was of 25 billion liters; while in 2015, it was around 129 billion liters. The world production of biodiesel grew 5.1 times between 2013 and 2015 [10, 11].

Diesel substitution for biodiesel, because of their similar characteristics, represents an alternative in many diesel applications, ranging from boilers to internal combustion engines [12].

It is essential to quantify the emissions caused by the CIE fueled by diesel and to determine the decrease in emissions regarding diesel. Biodiesel is acquiring increasing importance in the international context for it represents a rapid expansion of the industrial sector as a biofuel alternative to diesel [13, 14].

Currently, international governments and organizations are introducing new regulations that establish a stricter limit to emissions as an effort to mitigate the emissions of greenhouse gases [15–17]. To verify that a reduction of emissions or more efficient combustion is being obtained, it is imperative to have systems to quantify the emissions. Usually, the systems to register the internal combustion gas emissions use specialized analyzers that operate based on standardized methodologies [18]. These systems provide information about the concentration in parts-per-million (ppm) or concentration percentages according to the type of gas released in the combustion.

They are tools used to monitor and control combustion. At this time, there are many systems to determine the concentration of emissions; these systems can be portable or permanent, they present advantages and disadvantages, among these stands out the high acquisition cost with a range of 2000 up to 20,000 USD, rigidity, and the impossibility of being scalable [19, 20]. The use of these systems is limited to specific applications without having the versatility and flexibility to adapt them to other required uses. Nowadays, hybrid systems based on virtual instrumentations, a DAQ (data acquisition system), and a personal computer are being used as an alternative to the traditional systems customarily used to measure emissions.

The virtual instrumentation (VI) constitutes a new technology that covers the use of software and hardware systems that, with the use of a computer, replaces a measurement and control system in the real world. Any program and hardware that fulfill this function are a VI. In almost every commercial system, the concept of VI is realized in an object-oriented programming language [21]. The modern scientific instrumentation promotes the introduction, development, and evolution of VI-based systems. The main advantages of virtual instrumentation consist that they are defined by the end user, are scalable, recyclable and can connect with the outside world using modern communication technologies in addition to having a low cost per acquisition channel. In most cases, the VI has the possibility of modification, the facility of personalization to the specific necessities of each user, and the use of programming language [22]. The virtual instruments combine nonexclusive operation hardware with powerful software, obtaining a scalable architecture instrument, as a result, with the possibility of being modified if required [23–25].

Currently, the emission measurement of internal combustion gases originated from CIE is performed through the use of autonomous modular analyzers, dedicated and specialized, that provide the information about the quantity or concentration of the gases. In recent years, the application of virtual instrumentation for the measurement of said emissions has been proposed [26]. It is because of a virtual system allows to measure and monitor the CO concentration in vehicle exhaust gases that have been developed [27].

Recently, National Instruments developed a virtual instrument for the emission measurement generated by internal combustion engines. This instrument is based on the international emission standards, in particular, the Euro 4 and EPA. These agencies specify the total amount of pollutants that an internal combustion engine must emit to the atmosphere. These emission factor units are defined in general as gram per mile [28].

In the present work, biodiesel (B100) was obtained through soybean oil, with which were prepared mixtures with diesel: B2, B5, B10, and B20. The kinematic viscosity and heating value were determined. A virtual instrument for the measurement and monitoring of emission (VIEM) based on the virtual programming platform LabVIEW 2010® was developed.

The measurement of fuel consumption (FC), revolutions per minute (rpm), and exhaust temperature were realized based on the ARDUINO platform. The VIEM was synchronized with the sensors, data acquisition card, and the signal conditioners to measure and register in real time parameters like $O_2$, NO, CO, $SO_2$, $CO_2$, FC, rpm, and temperature. The VIEM programming, electrical schematic diagrams for the sensor signal conditioning, as well as the characterization of the $O_2$, NO, CO, $SO_2$, and $CO_2$ sensors are presented as results [29].

## 2. Materials and methods

### 2.1. Diesel-biodiesel mixtures

The fuels used were PEMEX diesel, ultra-low sulfur, and biodiesel. It was obtained biodiesel through the transesterification of soybean oil in the presence of methanol, by alkaline catalysis. With these fuels, the following mixtures were prepared: B2, B5, B10, and B20. Some of the physicochemical properties for each of the mixes are presented below.

### 2.2. Fuel kinematic viscosity

The determination of the kinematic viscosity of the fuel mixtures B2, B5, B10, B20, and diesel was performed using a CAP 2000+ Viscometer, where the information of the PEMEX diesel kinematic viscosity was entered, whose value is between 1.9 and 4.1 ($mm^2/s$) at 40°C [30]. All determinations were performed in triplicate.

### 2.3. Fuel heating value

The high heating value (HHV) measurements were performed by the utilization of a calorimetric bomb IKA WERKE model C2000 basic. The equipment was calibrated before experimentation. The procedure was followed according to ASTM E711. The tests were performed in triplicate to obtain reliable results.

### 2.4. VI for the measurement of $O_2$, CO, $CO_2$, NO, and $SO_2$ gases

The hardware system is composed of four systems. The first system is comprised of the $O_2$, NO, CO, $SO_2$, and $CO_2$ sensors listed in **Table 1**. The characteristics and operating ranges of each of the used sensor are described, which were of infrared response and work under the nondispersive infrared (NDIR) principle. This analyzer is used to detect the presence of carbon dioxide up to a volume of 100%. It is possible to determine said gas concentration through an infrared source with a specific filtration, which is assembled inside the optical/gas cavity. The infrared sensor interconnects with an electronic signal. The signal processing implies the linearization and compensation in temperature using algorithms in the system software. The infrared gas sensor uses a low frequency flash lamp drive that is controlled by an excitation circuit. Infrared radiation pulses reflect inside providing a trajectory through the gas and objective. "Pyros" pyroelectric detectors are used to determine the infrared signal change. The active pyroelectric is sensitive to the changes in the infrared wavelengths that are usually absorbed by the gas, passing between the transmitter and receiver [18].

Electrocatalytic analyzers utilize sensors that were developed as an outgrowth of fuel cell technology and are commonly used to detect $O_2$ concentrations in sample streams. These sensors use a solid catalytic electrolyte to aid the flow of electrons from a sample gas cell to a reference gas cell. In practice, catalyst-coated ceramic materials (such as $ZrO_2$) separate the reference cell (containing a high concentration of $O_2$) and the sample stream.

When heated, the electrolyte allows the transfer of ionic oxygen components from the reference cell to the sample cell. The surface of the electrolyte has a special electrocatalytic coating that catalyzes the transfer process and serves as an electrode surface to attract released electrons.

The ions are migrating from the reference side to the sample side release electrons on this surface, striving toward the equilibrium. However, since the sample is continuously replenished, a continual flow of electrons is induced across the measurement load resistor, and the current flow is used to infer the concentration of oxygen in the gas sample stream. Because the electrons contain a finite catalytic activity, it is necessary to establish a limit to the diffusion speed of objective gas in the sensor, guaranteeing that gas reacts appropriately. It is performed through a barrier taking the shape of a small hole or capillary located on the sensor cover [18].

| Features | Operating conditions | | | | |
|---|---|---|---|---|---|
| | $SO^*_2$ | NO | $O_2$ | CO | $CO_2$ |
| Operating ranges | 0–2000 ppm | 0–250 ppm | 0–30% $O_2$ | 0–500 ppm | 0–100% Vol. |
| Model | EC4-2000-$SO_2$ | EC4-250-NO | EC410 | EC4-500-CO | IR11BR |
| Sensitivity | 8–20 nA/ppm $SO_2$ | 320–480 nA/ppm of NO | N/A | 55–85 nA/ppm CO | N/A |
| Maximum overload | N/A | 1000 ppm | 100% $O_2$ | 1500 ppm CO | N/A |
| Zero in air at 20°C | <±50 ppm $SO_2$ | −0.06–4.5 ppm NO | N/A | <±3.1 ppm of CO | N/A |
| Zero deviation (−20–+40°C) | 0–4 ppm $SO_2$ | N/A | N/A | N/A | N/A |
| Resolution | 5 ppm $SO_2$ | 0.5 ppm NO | 0.1% $O_2$ | 1 ppm CO | N/A |
| response time | t 90 < 60 s | t 90 = 35 s | t 90 < 15 s | t 90 < 30 s | N/A |
| Temperature range | −20 to +50°C | −20 to +50°C | −20 to +55°C | −20 to +50°C | −20 to +55°C |
| Operating humidity | 15–90% RH Noncondensing | 15–90% RH Noncondensing | 15–95% RH | 15–90% RH Noncondensing | 0–99% Noncondensing |
| Pressure range | 90–110 kPa | 90–110 kPa | 90–110 kPa | 90–110 kPa | 80–120 kPa |
| DC supply recommended | N/A | N/A | N/A | N/A | 5 V |

**Table 1.** Sensors operating ranges.

The second system is composed of the conditioners and signal amplifiers for each one of the sensors. The third system is comprised of the DAQ card and PC. The function of the data acquisition card is to digitize the information provided by the signal conditioners.

The used DAQ was a USB model 6009, with eight analog inputs of 14 bits, 48 kS/s, two static analog outputs of 12 bits, 12 digital inputs-outputs, and a 32 bit counter of National Instruments. The digital signals are transferred to the PC model Sony Vaio VGN- CR190 Intel Core Duo T7100 @ 1.8 GHz, Windows 7 operating system.

### 2.5. Integrated circuits

The integrated circuits are used to condition and amplify the signal given by the sensors when the information about the gas quantity or concentration is registered. The voltage, Vo, registered is of 2.435 V when 1% of oxygen is present, being this the minimum concentration the sensor can detect. It means that the data acquiring card (DAQ) process increases equivalent to 0.065 V/1% of $O_2$. To determine the output voltage, in the case of the $O_2$ sensor's conditioning output signal, Eq. (1) was used.

$$Vo = 2.5\,V - 65\,mV/\% * [O_2\%] \quad (1)$$

In **Figure 1**, the integrated circuit diagram used to condition and amplify the signal provided by the $O_2$ sensor is presented.

The registered voltage, Vo, is of 2.508 V when 1 ppm of NO is present, being this the minimum concentration the sensor can detect.

It means that the data acquisition card DAQ process increases equivalent to 0.004 V/1 ppm of NO. Eq. (2) was used to determine the output voltage.

$$Vo = 2.5\text{ V} + 8\text{ mV/ppm}^{*}\ [\text{NO ppm}] \tag{2}$$

**Figure 2** presents the integrated circuit diagram used to condition and amplify the signal provided by the NO sensor.

The registered voltage, Vo, is of 2.5047 V when 5 ppm of $SO_2$ is present, being this the minimum concentration the sensor can detect. It means that the data acquisition card DAQ process increases equivalent to 0.00476 V/5 ppm of $SO_2$. For the output voltage estimation, Eq. (3) was used.

$$Vo = 2.5\text{ V} + 0.952\text{ mV/ppm}^{*}\ [SO_2\text{ ppm}] \tag{3}$$

In **Figure 3**, the integrated circuit diagram used to condition and amplify the signal provided by the $SO_2$ sensor is presented.

The registered voltage, Vo, is of 2.5039 V when 1 ppm of CO is present, which is the lower limit the sensor can detect. It means that the data acquiring card DAQ process increases equivalent to 0.00392 V/1 ppm of CO. For the output voltage calculation, Eq. (4) was implemented.

$$Vo = 2.5\text{ V} + 3.92\text{ mV/ppm}^{*}\ [\text{CO ppm}] \tag{4}$$

In **Figure 4**, the integrated circuit diagram used to condition and amplify the signal provided by the CO sensor is presented.

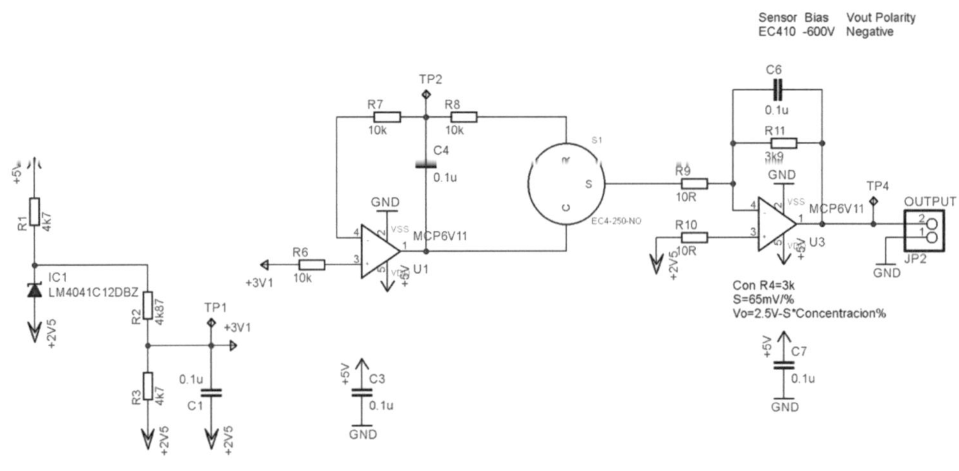


**Figure 1.** Designed circuit to condition the $O_2$ signal.

http://dx.doi.org/10.5772/intechopen.80533

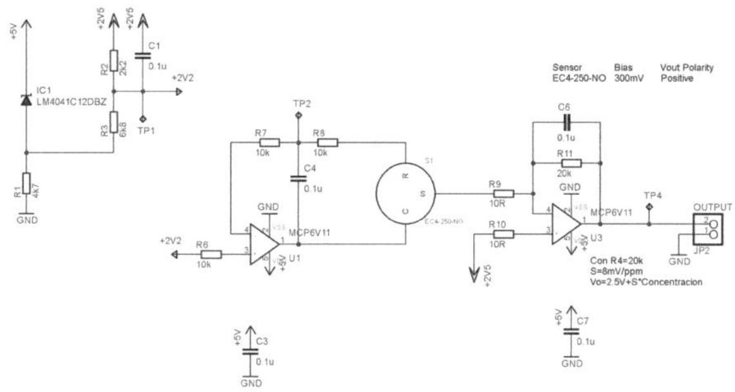


**Figure 2.** Designed circuit to condition the NO signal.

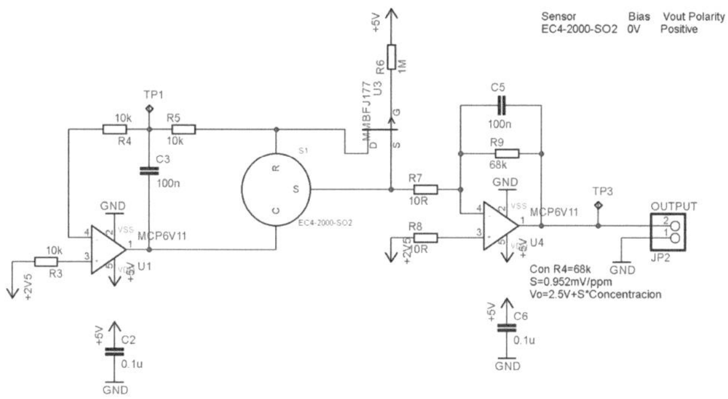


**Figure 3.** Designed circuit to condition the $SO_2$ signal.

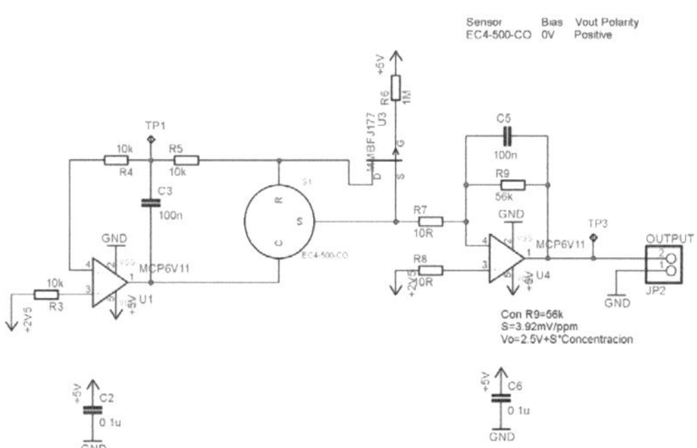


**Figure 4.** Designed circuit to condition the CO signal.

In **Figure 5**, the absorbance fraction regarding the $CO_2$ concentration is observed. It allows finding the sensor's sensitivity that ranges between 0 and 100% of concentration volume of carbon dioxide.

In **Figure 6**, the integrated circuit diagram used to condition and amplify the signal provided by the $CO_2$ is presented.

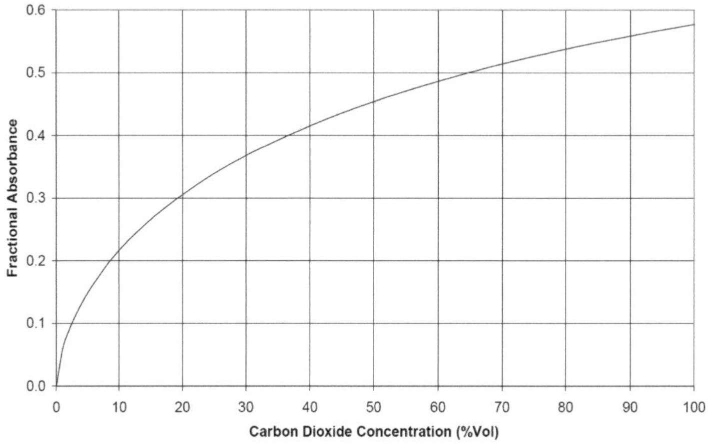


**Figure 5.** Absorbance's fraction relation regarding the $CO_2$ concentration.

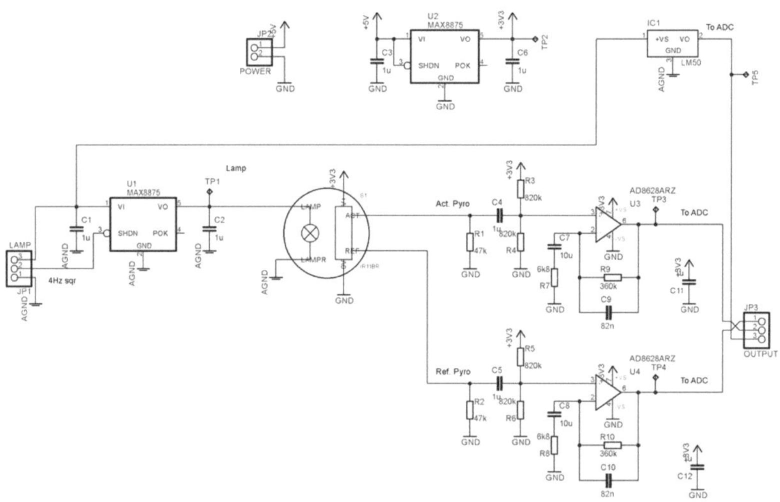


**Figure 6.** Designed circuit to condition the $CO_2$ signal.

## 2.6. Virtual instrument for emission measurement (VIEM)

This section shows the programming description of the main blocks that are part of the virtual instrument for emission measurement that was developed in a LabVIEW 2010 platform. In the figures, the operation performed by each programming block is shown. The VIEM counts the tests performed to organize the file experiment. It is made through the test counter (Sub-VI), which saves a text file (.txt) with a start value of zero and an increment of 1 for each VIEM implementation, in the "Configuration tab" through a "property node", and the menus and the toolbar are configured by the VIEM. **Figure 7** illustrates a programming block. Within while loop 1, routines are executed for test configuration and execution. The test duration, in turn, is selected through "Time test" control. The rhythm configuration of the sample is executed

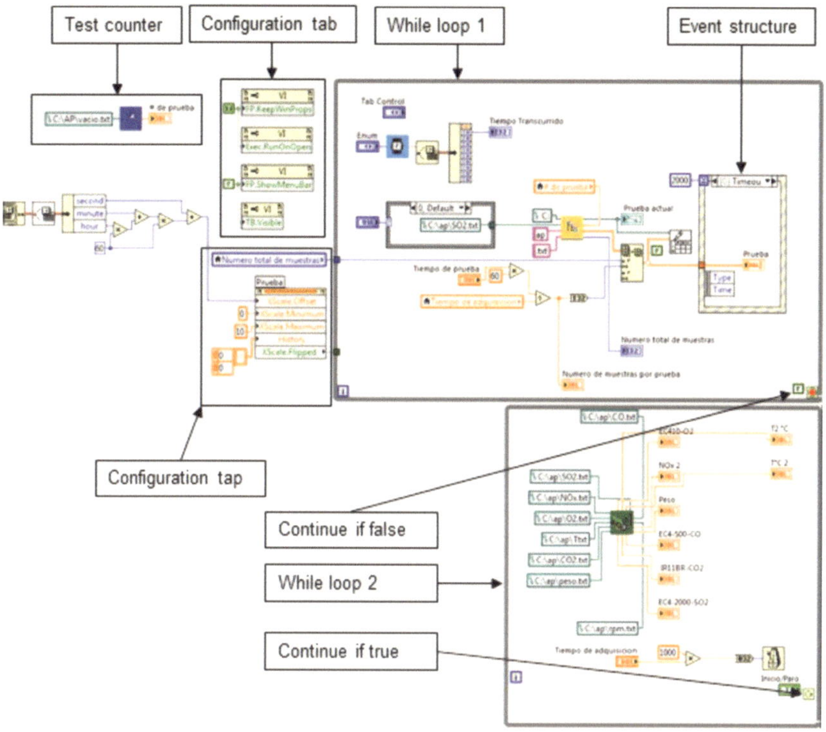


**Figure 7.** Block programming.

through "Acquisition time" control. The sub-VI "Direction by test" generates the name and the address of the file, which will contain the test samples in the implementation process.

The sub-IV "Write to Spreadsheet File" generates a new file for each registered variable which will be shown through the graphics display "Test." The latter is inside the "Event Structure" module, which allows synchronization with the computer clock through a timeout of 2000 ms, while loop 1 remains in continuous execution, while loop 2 is focused on the acquisition of sensor signals. It is possible because of the sub-VI "Acquisition." It also configures the entries of storage directions for the several variables to measure and as indicator outputs own numeric of the "double" type. The frequency of this cycle depends directly on the configured value in the "Acquisition time" control.

The cycle owns an implementation of the "Continue if true" type which is controlled through the "start/stop" button, and the sample of the program block is shown in **Figure 7**. The sub-VI "Acquisition" is focused on sensor data recompilation through the use of VI express "DAQ ASSISTANT," which is configured as an analogical entry, referenced to earth type (RSE) for continuous sample taking of voltage, in an allowed measurement range from −10 to 10 V. The output of VI "Acquisition" shows the measurements to the user through a numerical indicator. The data are processed for its storage using the sub-VI "Write to spreadsheet". All of this are made for a maximum amount of eight sensors, and the data flow is controlled by the "Flats sequence" structure to avoid information saturation in the acquisition time.

Graphical programming in LabVIEW 2010® uses the programming model by data flow unlike programming based on a text focused in a flow model control programming. Instead of laboriously write lines of text, with the risk of syntactic and logic errors, LabVIEW 2010® is based on the linked of icons with cables. Besides, to most of the users, including the experimented ones, it is easier to visualize and interpret graphical programming compared to text programming. The VIEM flowchart is displayed in **Figure 8**.

### 2.7. VI for temperature and revolutions per minute measurement

A virtual instrument was developed based on Arduino to register the run of the motor revolutions per minute using an infrared diode and a receiver diode. One of them is located in front of another with a distance no longer than 1.5 cm sending a continuous signal between the two: a "low" or a "0" V. Basically, the program quantifies the times that the signal was

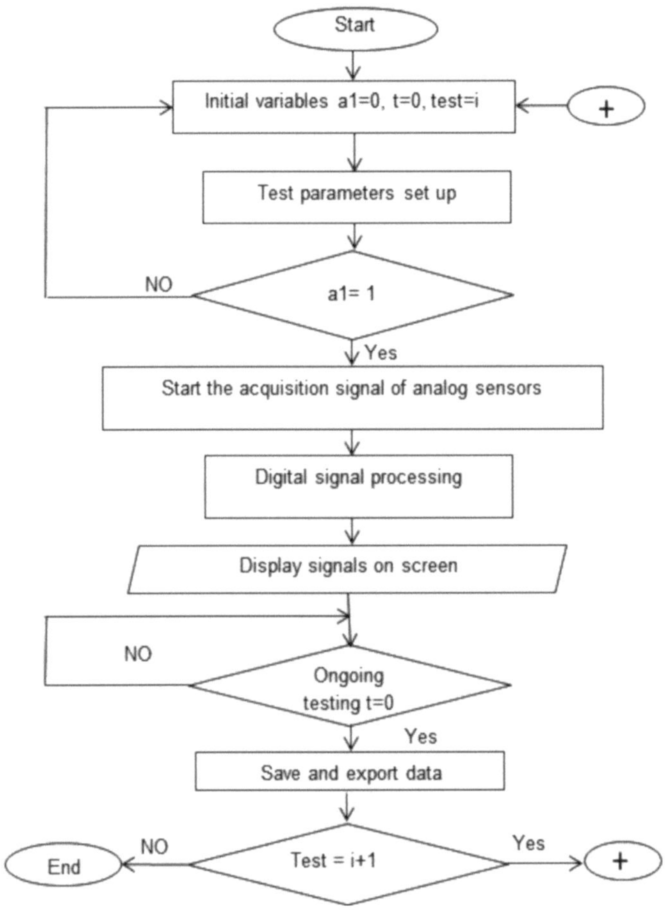


**Figure 8.** VIEM flow chart of the operation.

http://dx.doi.org/10.5772/intechopen.80533

interrupted, and it divides them in a minute obtaining rpm as a result. In **Figure 9**, it can be seen a block diagram describing the program operation based on the ARDUINO platform.

Once the program was obtained, the circuit will be welded on a printed circuit board. For temperature measurement, LM35 solid sensors were used which provide us a linear output signal with a scaling factor of 10 mV/°C, considering a range of −55 with a conversion factor of −550 mV up to 150°C with a conversion factor of 1500 mV, and this temperature range is enough to measure all the system temperature. A sensor was programmed in ARDUINO to compare the sensor precision, as it is explained in the block diagram of the program in **Figure 10**.

## 2.8. VI for measurement of fuel consumption

The used motor consumes 1.30 L/h in experimentation. Fuel consumption of the engine per second was quantified, relying on the result, and a virtual instrument was designed as well as

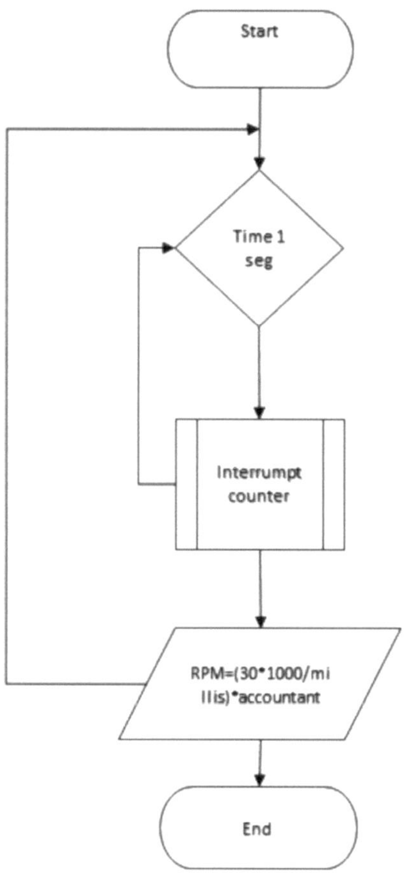


**Figure 9.** Block diagram of the program to measure rpm.

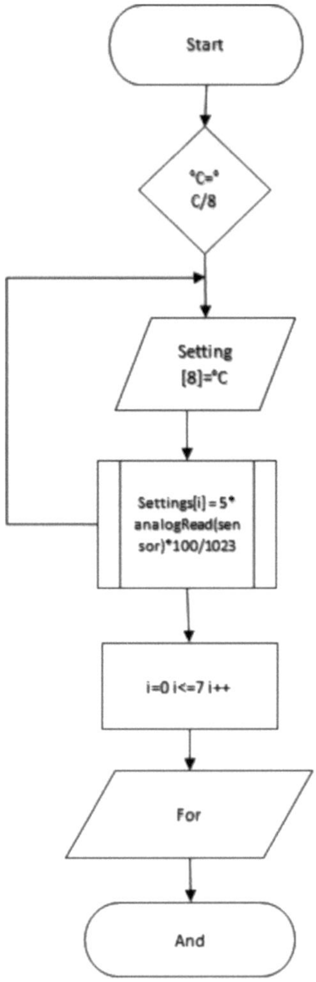


**Figure 10.** Programming operation in ARDUINO of the LM35 sensor.

the hardware. The VI uses a proximity sensor to measure the fuel difference and determine its consumption. The measurement is made by emitting and receiving ultrasonic signals with the proximity sensor. The proximity sensor is designed as an accessory for Arduino and integrates a signal conditioning system. **Figure 11** describes the operation of the virtual instrument with the use of Arduino as an interface for the measurement of fuel.

Tests were made to characterize the virtual instrument, using a measuring cylinder of 1 liter with a precision of ±10 mL. One of the benefits of this sensor is that it was not necessary to develop a signal conditioning circuit. It was programmed in ARDUINO. Based on the above, a 2.5 L capacity container was made using the acrylic material. The cubic container had a design area of 12.2 × 11.1 cm to obtain a total area of 135.4 $cm^2$. With the container area, a correction factor was added to the programming obtaining a final margin of error of ±50 mL.

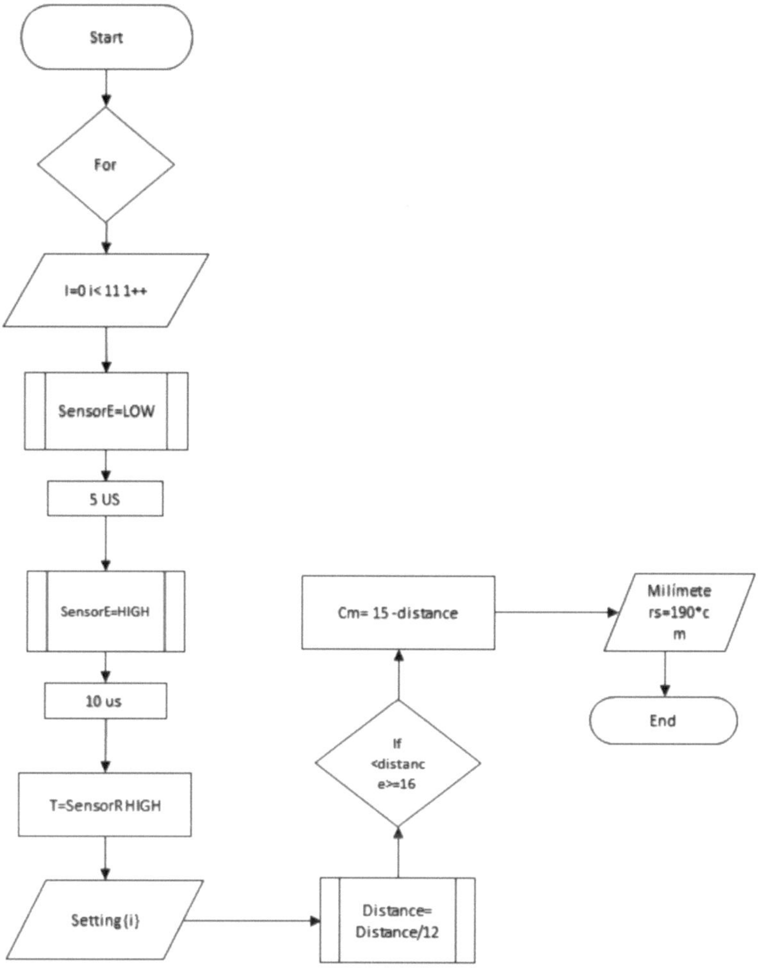


**Figure 11.** Block diagram of programming using Arduino.

## 3. Results

The viscosities of the different fuel mixtures in relation with temperature are listed in **Table 2**. The fact that the biodiesel viscosity is higher than the diesel one was observed. When improving the biodiesel percentage in the mixture, viscosity increases, which improves the lubrication of the parts that are in contact with the mix, such as the low pressure pump, the high pressure pump, injectors, and the system of fuel distribution [30–32].

About the high heating value determination of fuels, the diesel (PEMEX) presented a higher value of 44.97 MJ/kg and for the B100 was of 39.08 MK/kg. As a consequence, depending on the percentage of the biodiesel increase in the mixture, the high heating value decreases. The following values were obtained for the mixtures: B2 44.97 MJ/kg, B5 44.81 MJ/kg, B10 44.37 MJ/kg, and B20 43.68 MJ/kg [33].

| Temp (°C) | Viscosity (P) | Temp (°C) | Viscosity (P) | Temp (°C) | Viscosity (P) |
|---|---|---|---|---|---|
| **B2** | | **B5** | | **B10** | |
| 56.53 | 0.0180 | 49.90 | 0.0203 | 49.90 | 0.0203 |
| 59.13 | 0.0150 | 55.07 | 0.0187 | 55.10 | 0.0180 |
| 63.70 | 0.0140 | 59.63 | 0.0170 | 59.63 | 0.0160 |
| 67.50 | 0.0133 | 64.17 | 0.0160 | 64.20 | 0.0157 |
| 71.47 | 0.0130 | 68.77 | 0.0147 | 68.80 | 0.0137 |
| 75.43 | 0.0120 | 73.30 | 0.0143 | 73.30 | 0.0130 |
| 79.23 | 0.0110 | 77.90 | 0.0127 | 77.90 | 0.0120 |
| 83.33 | 0.0107 | 82.17 | 0.0123 | 82.13 | 0.0117 |
| 86.97 | 0.0103 | 87.00 | 0.0113 | 87.07 | 0.0110 |
| 91.37 | 0.0100 | 91.10 | 0.0113 | 91.10 | 0.0100 |
| **B20** | | **B100** | | **Diesel** | |
| 49.90 | 0.0213 | 54.57 | 0.0320 | 55.77 | 0.0210 |
| 55.00 | 0.0193 | 57.83 | 0.0303 | 59.37 | 0.0193 |
| 59.67 | 0.0177 | 62.37 | 0.0280 | 63.63 | 0.0173 |
| 64.10 | 0.0160 | 66.43 | 0.0257 | 67.57 | 0.0160 |
| 68.73 | 0.0143 | 70.57 | 0.0243 | 71.43 | 0.0150 |
| 73.33 | 0.0140 | 74.70 | 0.0227 | 75.43 | 0.0143 |
| 77.93 | 0.0127 | 78.80 | 0.0217 | 79.20 | 0.0133 |
| 82.13 | 0.0117 | 82.93 | 0.0200 | 83.40 | 0.0130 |
| 86.97 | 0.0110 | 87.03 | 0.0193 | 87.00 | 0.0123 |
| 91.13 | 0.0107 | 91.20 | 0.0183 | 91.17 | 0.0113 |

**Table 2.** Viscosity of the B2, B5, B20, B100 fuels and diesel.

The result of the virtual instrument development used to measure the CIE parameters is shown in **Figure 12**. The main user screen that exhibits the graphic interface represents the instrument when choosing the configuration option. The software allows defining the time test, the acquisition time, and the start option to proceed to the predefined tests.

**Figure 13** displays the user graphic interface, where the measures in real time of $O_2$, NO, CO, $SO_2$, and $CO_2$ can be observed, such as the temperature, the fuel consumption, and the rpm, which are being monitored and where information is being obtained from each one of the parameters considering the predefined values by the user.

The results obtained from the integrated circuits are specific, because signal conditioning circuits were designed for each sensor. It is unlikely that there are circuits that are the same

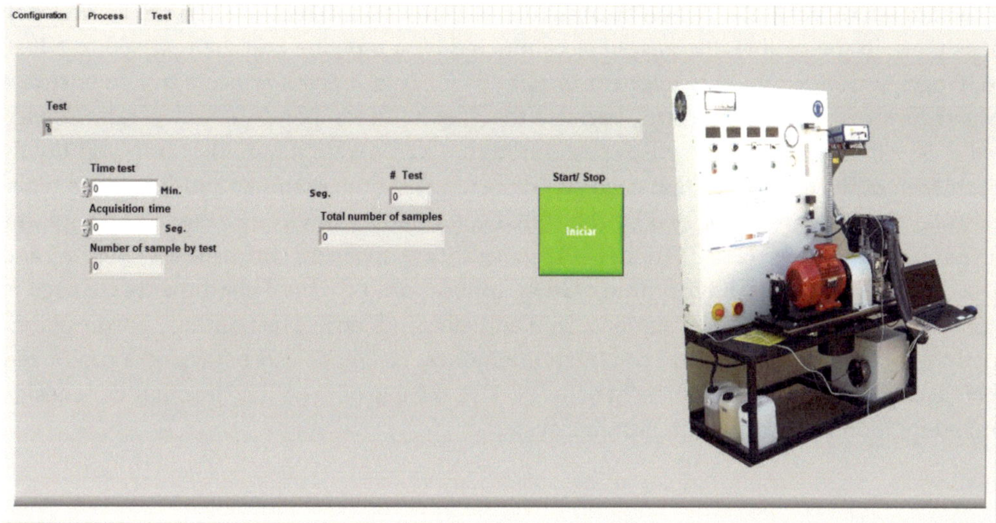


Figure 12. User graphic interface, principal window, and configuration tab.

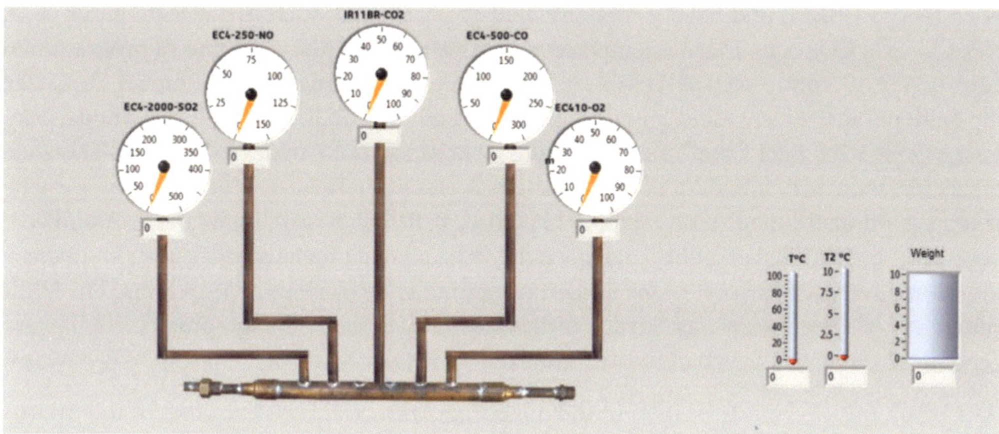


Figure 13. User graphic interface.

as those presented. It is not possible to access the programming of the other instruments sold commercially due to the protection of their programs.

## 4. Application

The VIEM main function is to measure the concentration of gases coming from the manifold of the combustion chamber of the CIE, where the $O_2$, NO, CO, $SO_2$, and $CO_2$ emissions of combustion gases are generated, by using infrared and electrochemical sensors. These sensors determine the gas concentration in ppm or percentage of the concentration of the emissions generated when burning diesel-biodiesel mixtures. When the combustion takes place in the engine, the gases are conducted by a pipeline and then passed to a heat exchanger to reduce the gas temperature at 40° C. This process is necessary to ensure a proper functioning of the sensors, because the maximum temperature for operation specified is 50 ± 5°C, after that the gases pass through a water trap, a humidity trap and finally reach each of the sensors so that each sensor reads real-time readings and generates registration data, databases provided by the signal conditioning devices processed through the virtual instrument on the PC. There is the versatility to measure the readings every 15 and 20 seconds, as well as 1 and 5 minutes. Other applications of virtual instruments are used in monitoring stations, chemical stations, iron and steel industries, restaurants, power plants, combined cycle, diesel internal combustion engines, gasoline, and natural gas motor test vehicle departments. They represent an option for the use of the VIEM through its redesign and adaptation to function as a portable system.

## 5. Conclusions

The development of hardware and software of the virtual instrument for the measurement of emissions (VIEM) and for the measurement of parameters such as emission gases of $O_2$, NO, CO, $SO_2$, $CO_2$, rpm, FC, and temperature is constituted by the platforms of programming LabVIEW 2010® and ARDUINO MEGA as well as a data acquisition card model DAQ 6009 the National Instruments and an acquisition card model ARDUINO ONE, PC model Sony Vaio VGN-CR190 Intel Core Duo T7100 @ 1.8 GHz, Windows 7 operating system. These elements make up a combination of nonexclusive operation hardware with powerful software, obtaining an instrument of scalable architecture, with the possibility of being modified if necessary, presenting versatility and flexibility to adapt to other required uses, showing as the main advantage the cost of the system compared with conventional systems. The VIEM shows in real time the results of each of the measured parameters, graphs the results, and generates a database of each of the experimental activity.

As a result of the characterization of the diesel-biodiesel mixtures, the value of the calorific value of the diesel that was used as a reference point is 44.97 MJ/kg, and for the B100, it is 39.08 MJ/kg. It is necessary to develop specific circuits for gas sensors such as $O_2$, NO, CO, $SO_2$, and $CO_2$, where the integrated circuits were designed to obtain equivalent increments of 0.065 V/1% of $O_2$, 0.004 V/1 ppm of NO, 0.00476 V/5 ppm of $SO_2$, and 0.00392 V/1 ppm of CO, with the schematic diagrams presented in this chapter and the model of the sensors, and it is possible to replicate the hardware used for gas measurement.

With the development of the VIEM, it is easier and accessible to implement a test bench used to quantify the performance, emissions, and efficiency of the CIE. It was obtained a more demanding control on the inputs and outputs of the system, as well as the operating parameters of the diesel engines using a biofuel, or any other engine that works with any other fuel, biofuel or mix of both considering the minimum and maximum parameters of the sensors.

## Acknowledgements

The authors thank the Consejo Nacional de Ciencia y Tecnología (CONACYT) and the Instituto de Ingeniería of the Universidad Autónoma de Baja California for its support in the development of this work. They also thank the Programa de Fortalecimiento de la Calidad Educativa 2018 (PFCE) for the financial support in publishing this chapter.

## Conflict of interest

The authors declare no conflict of interest.

## Author details

Armando Pérez[1], Rogelio Ramos[2], Gisela Montero[2*], Conrado García[2], Marcos Coronado[2], Héctor Campbell[2], René Delgado[1] and Alejandro Suástegui[3]

*Address all correspondence to: gmontero@uabc.edu.mx

1 School of Sciences of Engineering and Technology, Unit Valle de las Palmas, Autonomous University of Baja California, Tijuana, Baja California, México

2 Institute of Engineering, Autonomous University of Baja California, Mexicali, Baja California, México

3 Faculty of Engineering, Autonomous University of Baja California, Mexicali, Baja California, México

## References

[1] Environmental Protection Agency. Light-Duty Vehicle and Light-Duty Truck—Clean Fuel Fleet Exhaust Emission Standards [Internet]. 2017. Available from: http://epa.gov [Accessed: 14-11-2017]

[2] Plint M, Martyr A. Book Engine Testing: Theory and Practice. 4th ed. Oxford: Butterworth-Heinemann; 1995. pp. 1-549. DOI: 10.1016/B978-0-08-096949-7.00001-7

[3] Automotive Fuels—Fatty Acid Methylesters (FAME) for Diesel Engines—Requirements and Test Methods [Internet]. 2010. Available from: http://agrifuelsqcs-i.com/attachments/1598/en14214.pdf [Accessed: 23-02-2017]

[4] Ganjehkaviri A, Jaafar M, Hosseini S, Musthafa A. Performance evaluation of palm oil-based biodiesel combustion in an oil burner. Energies. 2016;**9**:9-97. DOI: 10.3390/en9020097

[5] Bosch R. Automotive Handbook. 6th ed. Wiley-Blackwell; 2004

[6] Daming H, Haining Z, Lin L. Biodiesel: An alternative to conventional fuel. Energy Procedia. 2012;**16**:1874-1885. DOI: 10.1016/j.egypro.2012.01.287

[7] Kurji H, Valera A, Okon A, Chong C. Combustion and emission performance of $CO_2$/$CH_4$/biodiesel and $CO_2$/$CH_4$/diesel blends in a Swirl Burner Generator. Energy Procedia. 2017;**142**:154-159. DOI: 10.1016/j.egypro.2017.12.025

[8] Montero G, Stoytcheva M, García C, Coronado M, Toscano L, Campbell H, Pérez A, Vázquez A. Current Status of Biodiesel Production in Baja California, Mexico. Rijeka: Intech; 2011. ISBN: 978-953-307-784-0

[9] Pérez A, Montero G, Ayala J, Coronado M, García C, Campbell H. Simulación en Aspen de la combustión de mezclas diésel-biodiesel. Ingeniería, Investigación y Tecnología. 2015;**16**:83-92. DOI: 10.1016/S1405-7743(15)72109-8

[10] Global Energy Transformation: A Roadmap to 2050 [Internet]. 2018. Available from: https://www.irena.org/publications/2018/Apr/Global-Energy-Transition-A-Roadmap-to-2050

[11] Lee H, Juan J, Taufiq Y, Kong P, Rahman N. Advancement in heterogeneous base catalyzed technology: An efficient production of biodiesel fuels. Journal of Renewable and Sustainable Energy. 2015:1-46. DOI: 10.1063/1.4919082

[12] Coronado M, Montero G, Valdez B, Stoytcheva M, Eliezer A, García C, Campbell H, Pérez A. Degradation of nitrile rubber fuel hose by biodiesel use. Energy. 2014;**68**:364-369. DOI: 10.1016/j.energy.2014.02.087

[13] Ngayihi C, Nzengwa R, Danwe R. Comparing in Cylinder Pressure Modelling of a DI Diesel Engine Fuelled on Alternative Fuel Using Two Tabulated Chemistry Approaches. Hindawi Publishing; 2014. p. 7. DOI: http://dx.doi.org/10.1155/2014/534953

[14] Lapuerta M, Rodríguez J, Agudelo J. Diesel particulate emissions from used cooking oil biodiesel. Bioresource Technology. 2008;**99**:731-740. DOI: 10.1016/j.biortech.2007.01.033

[15] Lin Y, Lee W, Wu T. Comparison of PAH and regulated harmful matter emissions from biodiesel blends and paraffinic fuel blends on engine accumulated mileage test. Fuel. 2006;**85**:2516-2523. DOI: 10.1016/j.fuel.2006.04.023

[16] McCormick R, Tennant C, Hayes R, Black S, Ireland J, McDaniel T, Williams A, Frailey M, Sharp C. In 2005 SAE Brasil Fuels and Lubicans Meeting; Mayo 2005; Rio de Janeiro, Brasil: NREL/CP-540-37508; 2005. pp. 1-11

[17] Norma Official Mexicana. nom-047-ecol-1993 [Internet]. 2010. Available from: http://www.diariooficial.gob.mx/normasOficiales.php?codp=5518&view=si#

[18] Clark N, Gautam M. Evaluation of Technology to Support a Heavy-Duty Diesel Vehicle Inspection and Maintenance Program [Internet]. 2001. Available from: http://www.arb.ca.gov [Accessed: 13-07-2017]

[19] Testo 340. Flue Gas Analyser [Internet]. 2018. Available from: http://www.testolimited.com/testo-340-industrial-flue-gas-analyser

[20] Enerac-Portable Combustion, Emissions & Gas Analyzer [Internet]. 2016. Available from: http://www.enerac.com/gas-analyzer-enerac-700/

[21] Meng H, Li J, Tang Y. Virtual instrument for determining rate constant of second-order reaction by pX based on labVIEW 8.0. Journal of Automated Methods & Management in Chemistry. 2009;**2009**:7. DOI: 10.1155/2009/849704

[22] Wang W, Li J, Wu Q. The design of a chemical virtual instrument based on LabVIEW for determining temperatures and pressures. Journal of Automated Methods & Management in Chemistry.2007;**2007**:7. DOI: 10.1155/2007/68143

[23] D'Ans G, Ripak P, Cerckel P, Degrève D, Katz M, Fontana P, Xheneumont A, Vanseven F. Some Applications of Virtual Instrumentation in NDT. St. Genesius Rode, Belgiu [Internet] 2017. Available from: http://www.ulb.ac.be/polytech/laborulb/athens/papere.pdf [Accessed: 11-12-2017]

[24] Booksh S. Virtual instrumentation. In Proceedings of National Technical Training Symposium and 26th Annual Pittsburgh; June 2002; Pennsylvania. USA

[25] Lenehan C, Barnett N, Lewis S. Design of LabVIEW-based software for the control of sequential injection analysis instrumentation for the determination of morphine. Journal of Automated Methods & Management in Chemistry. 2002;**24**:99-103. DOI: 10.1155/S1463924602000135

[26] Goldberg H. What is virtual instrumentation. IEEE Instrumentation and Measurement Magazine. 2000;**3**:10-13. DOI: 10.1109/5289.887453

[27] Draui PG, Ionel R, Gontean A-Ş, Ionel I. A new approach for carbon monoxide measurement using virtual instrumentation. Proceedings of The 6th WSEAS International Conference on Energy, Environment, Ecosystems and Sustainable Development (EEESD '10); October 2010; Timisoara, Romania. pp. 267-271

[28] Moore K. Testing Automotive Exhaust Emissions. National Instruments Application Notes. National Instruments Corporation; 2013. http://www.ni.com/pdf/app-note/testing-automotive-exhaust-emissions.pdf

[29] Branzila M. Instrumentation and Virtual Library for Air Pollution Monitoring. Rijeka: Intech; 2010. pp. 241-264. DOI: 10.5772/10051

[30] PEMEX. Hoja de datos de seguridad para el diesel [Internet]. 1998. Available from: http://www.pemex.com/negocio/gasolineras/nuestrosproductos/Documents/HDS-Pemex%20Diesel.pdf

[31] Phankosol S, Sudaprasert K, Lilitchan S, Aryusuk K, Krisnangkura K. Estimation of density of biodiesel. Energy & Fuels. 2014;**28**:4633-4641. DOI: 10.1021/ef501031z

[32] Pratas M, Freitas S, Oliveira M, Monteiro S, Lima A, Coutinho A. Biodiesel density: Experimental measurements and prediction models. Energy & Fuels. 2011;**25**:2333-2340. DOI: 10.1021/ef2002124

[33] Paniclelli L. Análisis de ciclo de vida (ACV) de la producción de biodiesel (B100) en Argentina [thesis]. Universidad de Buenos Aires; 2006. https://biodiesel.com.ar/download/acv_biodiesel_l_panichelli_tesis_uba_final.pdf

# Engineering Microbial Consortia for Bioconversion of Multisubstrate Biomass Streams to Biofuels

Fang Liu, Eric Monroe and Ryan W. Davis

Additional information is available at the end of the chapter

http://dx.doi.org/10.5772/intechopen.80534

**Abstract**

Production of biofuels from nonfood biomass has emerged as a sustainable option to address the problems associated with growing enery demand for transportation, heating, and industrial processes, in the context of diminishing petroleum reserves and global climate change. Biomass resources such as lignocellulose-rich biomass and microalgae, despite being abundant pose several challenges for efficient bioconversion to biofuels. Major challenges that must be addressed are the chemical complexity of the biomass and the associated feedstock variability. In this chapter, the role of microbial consortium-based biocatalysis strategies that are being developed to address these issues are reviewed and discussed. Microbial coculture biocatalysts are systems that are engineered to specialize in the conversion of a general class of substrates present in the biomass hydrolysates into biofuel intermediates, providing the capability of adapting to the variable composition of the feedstock. The techniques being developed to understand the interactions between the members of the bioconversion consortia and the corresponding population dynamics of the engineered cocultures are also discussed.

**Keywords:** biomass feedstocks, feedstock variability, biofuels, multisubstrate, microbial consortia, population dynamics

## 1. Introduction

The driving forces behind the urgent need for developing sustainable bioenergy and bioproducts include growing concerns over climate change, conflicts over oil supplies, and the desire to stimulate a new bioenergy economy [1]. Biomass is a renewable energy source that can offer a substitute for fossil-based transportation fuels and create economic opportunities and rural development while reducing climate impacts related to greenhouse gas emissions. Biomass is an energy resource derived from plant and algae-based materials that includes agricultural residues, forest resources, grasses, woody energy crops, algae, wet waste, municipal solid waste, urban wood waste, and food waste [2].

Utilization of industrial biotechnology for production of petroleum displacing products from low cost feedstocks faces major challenges because of recalcitrance of the biomass to hydrolysis as well as the high degree of biochemical complexity and variability in crude biomass hydrolysates [3]. In addition to unintentionally produced fermentation inhibitors from pretreatment, biomass hydrolysates contain a variety of carbohydrates, lignin, protein, lipids, organic acids, and ash, often at nearly stoichiometric ratios of each component. Although significant progress is underway for effective biomass pretreatment and utilization of each of the common types of fermentation substrates [4–6], the diversity of these substrates can result in spiraling costs for upstream separations, supplementation of fermentation media, serial fermentation steps, and product purification. From a technoeconomic perspective, any one of these unit operations can tip the scales against economic viability for a promising breakthrough. In terms of a biochemical or hybrid biorefinery, fractionation of the biomass to generate a spectrum of marketable products is assumed [7]; however, it is very rarely the case that the bulk of the biomass can provide significant or reliable economic return without upgrading. In light of this, process intensification based on single-pot bioconversion of multiple substrates provides a means for achieving dramatic improvements in reaction kinetics, yields, product titers, and separations while minimizing unit operations, thereby reducing cost and investment risk in chemical manufacturing [8]. Furthermore, consortium-based biocatalysis strategies provide the potential for utilizing ecological interactions, such as mutualism or commensalism, to increase biofuel production efficiency and yield [9, 10]. In this chapter, we will review promising biomass feedstocks and the associated feedstock variability challenges paired with the microbial consortia approaches that are being developed for bioconversion of multicomponent biomass hydrolysates to biofuels.

## 2. Biomass feedstocks for biofuel production

Although the current commercial biofuels industry is predominantly focused on feedstocks from harvestable components of food or feed crops (starch, sucrose, and oils), the potential of lignocellulose as an energy feedstock has received significant focus for second-generation biofuel production. Lignocellulosic biomass refers to the lignocellulose-rich materials available from terrestrial plants and crop residues, and is the most abundant renewable form of fixed carbon on earth. Lignocellulosic biomass is primarily composed of three polymers - cellulose, hemicellulose, and lignin. Cellulose, the major component of lignocellulosic biomass, is a linear homopolysaccharide consisting of 500–15,000 anhydrous glucose units that are linked by β-1,4-glycosidic bonds, with cellobiose as the smallest repetitive unit [11]. Hemicellulose is the second most abundant polymer whose role is to provide a linkage between lignin and cellulose. Hemicellulose is a short, highly branched polymer of pentose (e.g. D-xylose and L-arabinose) and hexose (e.g., D-mannose, D-galactose, and D-glucose) which forms a random and amorphous structure [12]. Lignin is the third major component of plant biomass in addition to cellulose and hemicellulose, accounting for 10–40% (w/w) of plant cell wall on the mass basis [13]. Lignin is an amorphous, randomly branched heteropolymer comprising of phenylpropanoid units [14]. Lignin covalently binds to side groups on neighboring hemicellulose units, forming a complex matrix that surrounds the cellulose micro-fibrils. As the dominant structural polymer in terrestrial biomass, for evolutionary reasons, the lignocellulosic biomaterial is highly recalcitrant to deconstruction and hydrolysis. Because of this recalcitrance, significant effort is required for pretreatment of the material using mechanical, chemical, physicochemical and biological methods or various combinations of these [15] to fractionate lignocellulose into its biochemical constituents which can be converted to biofuels and chemicals using biocatalysis. However, large scale adoption of lignocellulosic biofuels in the market has been slower than expected, and the bulk of lignocellulosic bioenergy production (~3% of the total energy consumed in the U.S.) has been restricted to industrial heat, steam, and electrical power instead of liquid fuels.

Microalgae are another potentially abundant biomass resource for biofuel production. Microalgae comprise a large and diverse group of uni- or multicellular phototrophic and heterotrophic organisms which, because of high growth rate, minimal competition for agriculture land use, the ability to utilize $CO_2$, and rapidly improve strains [16], microalgae may provide a dynamic addition to the sustainable fuels portfolio. Current markets for algae derived products have however been restricted to higher value products such as nutraceuticals, cosmetics, and foods. Microalgae have long been touted for their ability to accumulate large quantities of lipids. Oleaginous microalgae have oil contents of 20–50% by weight and several can exceed 80% oil by weight of dry biomass basis [17]; therefore, microalgae are being seriously considered as a feedstock for biodiesel production if the biomass production costs can be significantly reduced. Various species of microalgae produce a host of lipid subtypes, including mono-, di-, and triglycerides, phospholipids, glycolipids, waxes, carotenoids, hydrocarbons among others [18]. Microalgae are also rich in carbohydrates (5–23%) and proteins (6–52%) [19]; different species and cultivation conditions result in variable proportions of protein and carbohydrates. Some species, *Spirogyra* sp. for example, have high carbohydrate content, 33–64% [20], whereas the common nutraceutical algae strain, *Arthrospira*

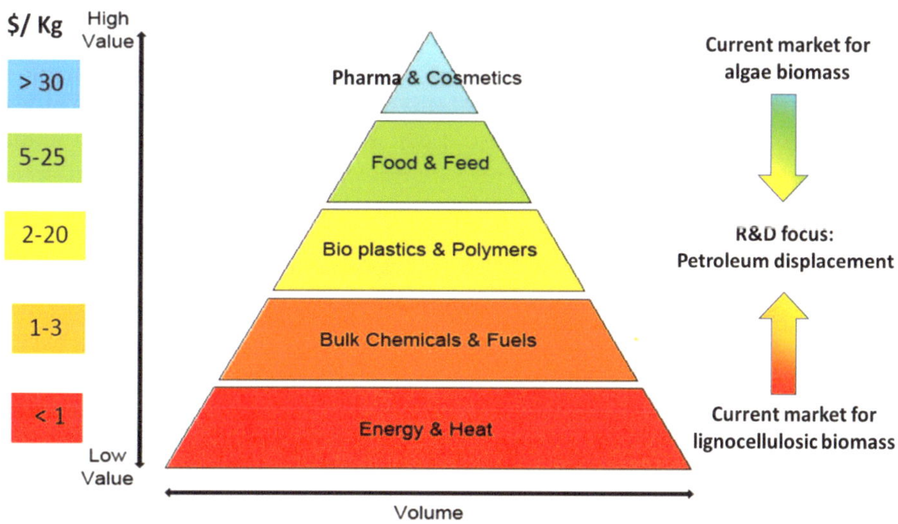


**Figure 1.** Current markets and the convergent R&D landscape for the general biomass feedstocks.

*platensis* (a.k.a *Spirulina* sp.) have has high protein content, ~67% [21]. During pretreatment of microalgae biomass for biofuels production, the carbohydrate and protein fractions of the microalgae feedstock can be hydrolyzed to soluble monomeric sugars and amino acids, each of which can be converted by a biocatalyst into biofuel products during a microbial fermentation process, thereby increasing the net yield of biofuel intermediates beyond that of the lipids alone. Although algae lack the extreme recalcitrance to pretreatment that is provided by lignin to terrestrial plants, the high biochemical diversity of algae presents similar challenges for consistent pretreatment. In both cases, however, efficient utilization of the bulk of the various biopolymers present in the biomass is vital for development of economically feasible bioconversion routes for biofuel production. **Figure 1** illustrates the current markets as well as the convergent R&D landscape for each of these general biomass feedstocks.

## 3. Variability of biomass feedstocks

One of the most important challenges facing consortium-based bioconversion technologies is feedstock variability. Changes in biomass composition, even within the same crop in the same geographic location, is an issue that has plagued many industrial bioprocessing plants and has greatly hindered the widespread application of bio-based renewable fuels and chemicals. Since many processing facilities require process optimization around a given biomass source for plant profitability, deviations in biomass composition often require the process to be re-optimized which can cause significant down time in the plant. While the impact on plant operation time has been reduced over time, feedstock variability remains one of the most significant cost drivers for many industrial bioprocessing operations.

While there are many biomass composition variables such as ash and moisture content that are important economic drivers for the larger biorefinery picture, most relevant to the bioconversion process is the biochemical breakdown of the biomass source. **Table 1** highlights the variability of the glucan versus xylan percentages for different crops. The large range of values represents the severity in variation which is most dramatic in corn stover, with a maximum of 66% and a minimum of 39% [22].

One technique that has been deployed to curb these issues is blending of different biomass sources to help with geographic and seasonal variability [1]. This strategy can reduce the impact of feedstock variability, but requires sophisticated biomass logistics management to

| | Corn stover | Miscanthus | Wheat |
|---|---|---|---|
| Mean | 53 | 61 | 50 |
| Maximum | 66 | 65 | 57 |
| Minimum | 39 | 55 | 41 |
| Range | 27 | 10 | 15 |

Data from [22].

**Table 1.** Carbohydrate compositions (% xylan + glycan) for various crops.

prioritize variables for maximum operational profitability and is intrinsically limited in its reach due to geographic and seasonal constraints. This strategy is further limited by biomass storage challenges that limit the availability of different sources for blending.

Differences in biochemical composition of various biomass sources have been well documented in the technical literature because biochemical conversion routes often utilize only specific components of the biomass (such as glucose and xylose) and large variations of these components will have major impacts on process efficiency and yields. The majority of this work has focused on the relative biochemical fractions of a particular biomass source that can be classified as carbohydrates, proteins, or lipids, commonly referred to as the proximate composition. In the bioconversion context, these three major biochemical classes have varying importance for different biomass sources, with carbohydrates receiving particular attention for most fermentation-based processes. These variations are explored for three biomass sources with near-term industrial relevance in the section below.

### 3.1. Corn stover

Ethanol production from corn stover has been a significant area of research and development due to the widespread availability of the feedstock from cultivation of maize. The updated Billion Ton Study released by the U.S. Department of Energy in 2016 reports that the base case for corn stover supplied under $80/dry ton is over 129 million dry tons per year of biomass [1]. The variation of the composition of corn stover from cultivation in different geographic locations is highlighted in **Table 2** [23].

### 3.2. Distillers grains with solubles

Distillers grains with solubles (DGS) is another example of a highly variable biomass source that may play an important role in scale-up of bioconversion technologies. Twenty-three million tons of DGS were produced in 2017 as a coproduct of the corn ethanol process, with significant quantities of similar coproducts such as wet grains and dried grains without solubles also being produced. Because DGS are a coproduct of corn-ethanol fermentation, they suffer from variability of the original crop used as well as differences in different fermentation batches and strategies. These variations are highlighted in **Table 3** [24–28].

| | South Carolina | Central Alabama | Northern Alabama |
|---|---|---|---|
| Cellulose | 41.13 (0.98) | 41.76 (0.18) | 41.22 (0.18) |
| Hemicellulose | 31.97 (0.99) | 21.98 (0.14) | 22.72 (0.16) |
| Lignin | 5.97 (0.48) | 6.49 (0.06) | 6.56 (0.08) |
| Holocellulose | 73.1 (1.42) | 63.74 (0.17) | 63.94 (0.15) |
| Ash | 3.97 (0.14) | 2.87 (0.02) | 3.43 (0.04) |

Data sourced from [23].

**Table 2.** Relative % (standard error) for corn stover harvests in the southeastern united states at three different locations.

### 3.3. Microalgae

While microalgae research has generally focused on high lipid strains and cultivation strategies for oil extraction and biodiesel production, recent studies have demonstrated the techno-economic necessity for higher productivity algal systems which generally corresponds to lower quantities of lipids and a higher fraction of proteins and carbohydrates [29]. Because of the wide diversity of potential strains and cultivation conditions, microalgal biomass has a wide range of biochemical possibilities. This is further magnified by the distinct difference in biomass composition at different stages in microalgal cultures growth cycle. As a part of a study to understand the impact of this variation on a dilute acid hydrolysis pretreatment process, the composition of three microalgae strains at different points in their growth cycle was investigated by Peinkos et al. and is summarized in **Table 4** [27].

The challenges caused by feedstock variability provide a unique opportunity for consortium bioconversion strategies. For many biomass sources, techno-economic reports conclude that the cost of the biomass itself is the main economic sustainability driver [29], so a growing field is developing around the use of consortium conversion strategies that enable utilization of the vast majority of the biomass for production of a biofuel or petroleum displacing commodity chemical in an integrated biorefinery scheme. Clearly, if the biomass itself is the key cost driver at scale, then as much value as possible needs to be extracted from of the biomass. Consortium strategies allow for different organisms to specialize in efficient conversion of a particular substrate and collectively convert as much of the diverse and variable biomass-derived biochemical intermediates to a fuel or chemical product as possible. If done effectively, this strategy will

| | Spiehs and Whitney [28] | Liu [26] | Belyea et al. [25] | Cromwell and Herkelman [27] |
|---|---|---|---|---|
| Protein | 30.2 (2.9) | 27.4 (3.3) | 31.4 (2.5) | 29.7 (5.7) |
| Oil | 10.9 (1.2) | 11.7 (2.2) | 12.0 (1.7) | 10.7 (5.0) |
| starch | — | 4.9 (2.5) | 5.3 (0.7) | — |
| Total carbohydrate | 53.1 | 56.5 | 52.1 | 54.3 |
| Crude fiber | 8.8 (1.4) | — | 10.2 (1.0) | — |

**Table 3.** Biochemical composition of DGS as a dry mass % (range of values) from four different studies as reported by [24].

| Species | Growth stage | Lipids | Ash | Carbohydrate | Protein |
|---|---|---|---|---|---|
| *N. granulata* | Early | 12.28 (0.16) | 14.2 | 8.92 (0.13) | 32.7 |
| | Mid | 25.6 (0.20) | 13.6 | 11.12 (0.48) | 23.1 |
| | Late | 57.33 (0.09) | 5.1 | 10.89 (0.11) | 9.4 |
| *C. vulgaris* | Early | 12.07 (0.09) | 6.7 | 11.12 (0.12) | 43.2 |
| | Mid | 15.02 (0.16) | 4.4 | 35.69 (0.01) | 24.0 |
| | Late | 23.14 (0.19) | 5.3 | 38.00 (0.36) | 15.2 |

Data from [27].

**Table 4.** Biochemical composition in mass of dry cell weight (Std Dev) for two microalgae species.

be an effective tool to reduce the impact of feedstock variability, as the ratio of growth of the different specialized organisms in the consortium can be engineered to be proportional to the concentration of the different substrates derived from a given biomass source.

## 4. Microbial consortia for biomass conversion and biofuel/chemical production

Recent advances in synthetic biology, metabolic engineering, and systems biology have enabled rapid progress in developing microbial cell factories [6, 30, 31] and novel enzyme cascade systems [32–34] for the conversion of biomass feedstocks and synthesis of biofuels and other platform chemicals. Although there are some successful examples of developing 'superbugs' capable of multiple functions, engineering a single microbe to simultaneously perform multiple tasks is still quite challenging and bioenergetically costly under most situations. Because of the complexity and the multisubstrate nature of the biomass feedstocks, it is especially challenging to engineer a single microbe to efficiently convert the diverse substrates (carbohydrates, proteins, fatty acids, oils, etc.) of the biomass to produce value-added products. In contrast to the 'superbug' paradigm, in nature microbes rarely live in isolation, but rather exist in highly diverse and complex communities known as consortia. The microbes in these communities interact in numerous ways ranging from cooperation to competition and are often capable of performing tasks that are far too complex for any single organism to complete themselves [35]. Besides the ability to perform complex biosynthetic tasks, microbial consortia exhibit many other appealing properties such as stability, productivity and functional robustness. Inspired by the powerful features of the natural consortia, there are rapidly growing efforts been undertaken to understand natural consortia and to engineer synthetic consortia for biotechnology applications [36, 37]. Well-designed microbial consortia involving two or more microbes can take advantage of the functions of individual microbes and their interactions to realize synergistic division of labor and more efficient utilization of biochemical substrates than monocultures. Natural and synthetic microbial consortia developed for the conversion of different biomass feedstocks for biofuel and chemical production will be discussed below.

### 4.1. Anaerobic digestion

When considering consortium-based bioconversion technologies, anaerobic digestion offers a model process due to its long history of industrial application of a complex, albeit unsupervised, microbial consortium. Anaerobic digestion is a biochemical process that converts organic material into a mixture of methane and $CO_2$ (biogas) in an anaerobic environment. Its widespread use since the middle of the nineteenth century and the extensive research around the details of the process make it an important benchmark technology as well a valuable resource for future consortium conversion technologies [38]. The anaerobic digestion process is generally split into five stages, creating an interconnected web of processes each utilizing naturally adapted microorganisms that play a critical role in the overall conversion process. The stages can generally described as disintegration, hydrolysis, acidogenesis, acetogenesis, and methanogenesis which describes the sequential process of the organic polymers comprising biomass (such as carbohydrates and proteins) being disintegrated, those polymers being hydrolyzed to their corresponding monomers, the monomers being fermented to organic acids and hydrogen, the long carbon chain organic acids being further broken down into acetic acid and hydrogen, and finally the conversion of acetic acid to the fully reduced methane product.

Anaerobic digestion offers a touchstone example of the technological potential of consortium bioconversion processes because many microorganisms involved are strongly dependent on other organisms in the consortium. This is exemplified by the interspecific transfer of hydrogen, i.e. hydrogenotrophs utilize the hydrogen being produced by syntrophic organisms oxidizing organic acids that require low pressures of $H_2$ [39]. These syntrophic organisms are not able to grow in pure cultures because of this interdependency. While these symbiotic relationships are in many ways beneficial, they leave the entire process vulnerable to varying environmental conditions. If the temperature, pH, or another important environmental factor becomes unfavorable to an important organism in the anaerobic consortium the whole process is stalled and operates less efficiently. Furthermore, because of the wide variety of microorganisms involved in the anaerobic digestion process and the high level at which conditions can be controlled, it is difficult to engineer the system for any organism or class of organisms. The diversity of microbes involved in anaerobic digestion has minimized the potential roles for including metabolic engineering into the process, with most recent works simply looking to understand the population dynamics of the organisms involved [40]. The robustness of the technology makes it a base case bioconversion route for almost any biomass source, many of which have been investigated to varying levels of detail. Each biomass source presents unique challenges and difficulties to the overall anaerobic digestion process; microalgal biomass is a key example, since cells have been observed to remain intact after a 30 day retention time in a digester [41].

### 4.2. Consolidated bioprocessing of lignocellulosic biomass

Microbial conversion of lignocellulosic biomass requires multiple biological functionalities, including production of saccharifying enzymes, enzymatic hydrolysis of lignocellulose to soluble fermentable sugars, and metabolism of sugars to desired biofuel products. Cellulose and hemicellulose in lignocellulosic biomass can be hydrolyzed by cellulase and hemicellulase enzymes such as endoglucanase, exoglucanase, β-glucosidase, xylanase, xylosidase, and arabinofuranosidase. Microorganisms that are able to hydrolyze lignocellulosic biomass or secret cellulolytic hydrolysate enzymes are important for this step. Most biological process employed fungi such as *Trichoderma reesei*, *Penicillium echinulatum*, *Penicillium purpurogenum*, *Aspergillus niger*, and *Aspergillus fumigatus* [4] which secrete cellulolytic enzymes for saccharification. Some anaerobic bacteria such as *Clostridium thermocellum*, *Ruminococcus flavefaciens* and *Clostridium cellulovorans* developed large, extracellular multienzyme complexes called the cellulosomes, which are highly structured and consist of multiple cellulolytic enzyme units that interact with each other synergistically [42]. These anaerobic bacteria are also found to degrade cellulosic substrates efficiently. To reduce the number of unit operations and capital costs, a nascent approach called consolidated bioprocessing (CBP) has been developed for simultaneous enzyme production, hydrolysis, and fermentation by employing the microbial consortia biocatalyst strategy.

The microbial consortia for CBP involve a cellulolytic strain that hydrolyzes hemicellulosic biomass to fermentable sugars and a fermentation strain that utilizes the cellulosic sugars for growth and conversion to biofuel products through the natural or engineered metabolic pathways. Native cellulosic microorganisms such as fungi and some anaerobic bacteria are usually selected as the cellulosic strains in the consortia. For example, the thermophilic anaerobe *Clostridium thermocellum* which is capable of hydrolyzing cellulose through the activity of the cellulosome multiprotein complex was cocultured with the engineered ethanol producing strain *Thermoanaerobacterium saccharolyticum* for the production of ethanol from Avicel [43]. By genetically engineering the two strains to be acetate and lactic acids-deficient, the yield was greatly improved, resulting in the coculture producing 38 g/L ethanol from 92 g/L Avicel. In another study, Minty et al. designed a synthetic fungal/bacterial consortium that involved a cellulolytic fungus *Trichoderma reesei* and an engineered *E. coli* strain for bioconversion of lignocellulosic feedstocks [44]. In the consortium, the cellulolytic fungus secreted cellulase enzymes to hydrolyze pretreated corn stover into fermentable sugars and the *E. coli* converted soluble saccharides into isobutanol with titers up to 1.88 g/L. The *T. reesei*/*E. coli* consortia developed in this study showed a cooperator-cheater dynamics which lead to stable equilibrium population.

The main challenges of using native cellulolytic microorganisms as the cellulolytic strains are the availability of tools for genetic manipulation and the application of these tools to engineer the strains with high yield, titer and robustness under industrial conditions. Industrial strains that are not naturally capable of hydrolyzing cellulose have also been genetically engineered as the cellulolytic strains. For example, *E. coli* was engineered to express cellulase, xylanase, β-glucosidase and xylobiosidase enzymes extracellularly [45]. The resulting two *E. coli* strains were able to grow on cellulose and xylan substrates, respectively. Subsequent incorporation of biofuel production pathways into each cellulolytic and hemicellulolytic strain enabled production of biofuel products such as fatty-acid ethyl ester, butanol and pinene from ionic liquid pretreated switchgrass by the *E. coli* coculture. In this *E. coli* based CBP, cellulose and hemicellulose in the pretreated biomass were hydrolyzed into soluble oligosaccharides by the secreted cellulases and hemicellulases from the two *E. coli* strains. The oligosaccharides were further converted to monomeric sugars by the expressed β-glucosidase enzymes and were eventually metabolized into biofuels via the heterologous pathways of the same *E. coli* strains. The *E. coli* strains developed in this study incorporated the biomass-degrading capability and biofuel-producing ability into one strain, unlike in most typical CBP systems where cellulolytic and fermentation strains are engineered separately. Instead, this *E. coli* coculture divided the labor of producing advanced biofuels from cellulose and hemicellulose fractions of the lignocellulosic biomass between the two engineered strains.

Besides biofuels, valuable chemicals are also produced from nonfood agricultural resources by the CBP approach. For example, Bayer et al. used a synthetic metagenomics approach to identify methyl halide transferase (MHT) genes from various organisms and screened these enzymes in *E. coli* for MHT activity [46]. MHT with the highest activity was cloned in *Saccharomyces cerevisiae* and the resulting strain was able to produce methyl halides from monomeric sugars such as glucose and xylose. The production of methyl halide from unprocessed switchgrass, corn stover, sugar cane bagasse, and poplar was then demonstrated by the consortium of the engineered yeast and a cellulolytic bacterium *Actinotalea fermentans*. In the coculture, the cellulolytic bacterium produced monomeric sugars from hemicellulosic biomass which were further converted to methyl halide by the engineered yeast.

Inspired by the natural cellulosome in some anaerobic bacteria for the hydrolysis of cellulose, Tsai et al. developed a synthetic yeast consortium displaying an artificial minicellulosome for the synergistic saccharification and fermentation of cellulose to ethanol [47]. In the consortium, one yeast strain displayed a miniscaffolding consisting of three different cohesin domains and three other yeast strains secreted dockerin-tagged endoglucanase, exoglucanase, and β-glucosidase, respectively. Because of the specific interaction between the orthogonal cohesin-dockerin pairs, dockerin-tagged cellulolytic enzymes bound onto the scaffolding and a minicellulosome structure formed on the yeast cell surface. Cellulose was hydrolyzed to glucose by the three cellulases in the yeast displayed minicellulosome and glucose was then utilized by the yeast and fermented to ethanol. In this work, it was found that the ethanol yield could be fine-tuned by adjusting the ratio of the different yeast strains in the consortium. The consortium with the optimized ratio of the different populations produced ethanol up to 93% of the theoretical value from cellulose.

### 4.3. Microbial consortia for variable sugar mixture feeds

Hydrolysis of lignocellulosic biomass by chemical or enzymatic methods usually results in a mixture of $C_6$ sugars (glucose, mannose, and galactose) and $C_5$ sugars (xylose and arabinose) [15, 48]. The relative proportion of these sugars varies among different hemicellulosic biomass sources. Although glucose is the most abundant hexose and xylose is typically the major pentose, the fraction of other sugars found in hydrolysates can also be significant depending on the biomass and the pretreatment process. Natural microbes commonly consume hexose sugars prior to initiating the consumption of a pentose sugar. This phenomenon is known as carbon catabolite repression, in which the presence of a preferred substrate represses the expression of genes in the microbes required for the metabolism of other substrates [49]. Although efforts have been made to develop a single organism that can consume glucose and xylose simultaneously [50–52], sugar mixtures remain inefficiently consumed in these single-organism processes. Moreover, since the sugar composition and proportion vary in different biomass hydrolysates, a single organism has very limited ability to adjust to this variation. Instead, multiorganism coculture systems with each organism selectively consuming only one particular sugar substrate have been developed to eliminate the carbon catabolite repression and convert the mixed sugar feed streams more efficiently and completely. The multiorganism systems also have the ability to adjust the ratio of the different individual populations in the coculture in order to adapt to the fluctuation in the mixed sugar compositions of the feedstocks.

As a recent example, two substrate-selective strains of *E. coli* were metabolically engineered with one strain only able to consume glucose and the other one only able to utilize xylose [53]. To construct the xylose selective (glucose deficient) strain, three genes *ptsG*, *manZ* and *glk* involved in glucose uptake were knocked out from the chromosome and the glucose-selective (xylose deficient) strain has the *xylA* gene encoding for xylose isomerase deleted. This coculture was demonstrated for simultaneous conversion of glucose and xylose for acetate production. Moreover, the relative biocatalyst concentrations in the coculture could be adjusted in order

http://dx.doi.org/10.5772/intechopen.80534

to optimize the product yield from feed streams with variable sugar compositions. Xia et al. extended the consortium strategy for mixed sugar utilization to a synthetic mixture composed of three sugars-glucose, xylose and arabinose as well as the growth inhibitor acetic acid [54]. In this case one strain engineered to utilize acetate but not sugar was used to selectively remove acetate in the first stage of the bioprocess. In the subsequent stage, three *E. coli* strains which were each engineered to utilize only one sugar were cocultured and consumed glucose, xylose and arabinose simultaneously. Although reduction of the net growth rate of the strains was observed as a result of the chromosomal gene deletions that were required to generate the strains, the consortium showed significant improvement in multiple sugar utilization than any single-organism approach.

Substrate-selective consortia have also been engineered for the production of biofuels from hemicellulose-derived sugar mixtures. In a recent study, the synthetic pathway of n-butanol was engineered into a glucose-selective strain and a xylose-selective strain and the resulting *E. coli* coculture produced n-butanol from the sugar mixture [55]. The system was further improved by distributing the n-butanol production pathway into the two strains for improved redox balance and 5.2 g/L n-butanol was produced by the coculture from a glucose-xylose mixture. In another related study, *Saccharomyces cerevisiae* was genetically engineered to utilize L-arabinose. By coculturing the L-arabinose-utilizing yeast and the yeast that could ferment glucose and xylose, the hexose and pentose sugars present in corn stover hydrolysates was efficiently fermented to ethanol [56].

An *E. coli* coculture system with two substrate-selective strains has also been developed to produce commodity chemical such as *cis, cis*-muconic acid from a glucose/xylose sugar mixture [57]. To overcome the high-level intermediate accumulation during the synthesis, the entire muconic acid synthetic pathway was split into the two *E. coli* cells. Therefore, one strain was only able to utilize xylose and convert it to the intermediate 3-dehydroshikimic acid (DHS) which is the precursor for muconic acid. The second strain incorporated a DHS transporter and could utilize glucose and convert DHS to the final product muconic acid through the rest of the split pathway. The *E. coli-E. coli* coculture system has shown to use sugar mixtures efficiently even at increased scale, and produce significantly higher amount of muconic acid than previous studies, with a yield of 51% of the theoretical maximum.

### 4.4. Microbial consortia for sugar and protein mixture

Besides sugar mixtures, many biomass resources are also rich in proteins. For example, distillers grains with solubles (DGS) produced from the first-generation bioethanol process are considered as a rich source of cellulosic polysaccharides (52–57%) and proteins (27–31%) [58]. Microalgae are another example of the biomass resource that is rich in polysaccharides and proteins. Therefore, developing biochemical techniques for simultaneous utilization of the sugar and protein fractions of the biomass is important for efficient bioconversion of these biomass feedstocks.

Microbes normally cannot utilize amino acids as carbon source for growth. However, an *E. coli* strain that was able to use 13 individual amino acids as the sole carbon source for growth was developed after several rounds of chemical mutagenesis. A strategy was subsequently developed to deaminate amino acids and convert the remaining carbon backbones of amino acids to fusel alcohols through metabolic engineering [59]. In a follow-on study, the engineered *E. coli* was further improved by modifying the cofactor specificity of two enzymes involved in the metabolic pathway and the resulting strain can produce fusel alcohols with significantly improved yield [60].

Liu et al. has demonstrated the feasibility of one-pot bioconversion of the protein and carbohydrate fractions of the DGS hydrolysates and algae hydrolysates into mixed fusel alcohols by an *E. coli* coculture [61]. In the consortia, one strain is dedicated for bioconversion of hexose and pentose sugars in the hydrolysates into isobutanol (C4) and isopentanol (C5) fusel alcohols, whereas the other strain was engineered to convert free amino acids into mixed fusel alcohols. At the optimized inoculation ratio of the two strains, the consortium produced the highest titer of total fusel alcohols, up to 10.3 g/L, including 6.5 g/L isobutanol which comprised 63.1% of the total alcohols. Correspondingly, the consortium with the optimized inoculation ratio consumed the highest amount of DGS carbohydrates and proteins in the hydrolysates, including near complete consumption of the glucose and arabinose and 85.1% of the xylose, as well as 31.3% of the total proteins in the hydrolysates. Evaluation of the biofuel properties of the fusel alcohols produced using this strategy indicates that the higher carbon chain length alcohol mixture (especially C3-C5) provides increased energy densities and a variety of improved physical properties, such as reduced water solubility and corrosivity, than ethanol [16]. Therefore, the mixed fusel alcohols produced in this coculture have promising potential applications as a fuel upgrading feedstock in gasoline, diesel, and jet fuel or as a neat fuel of itself.

Similar *E. coli* cocultures were also developed for the production of terpene mixtures from the carbohydrate and protein fractions of algae hydrolysates [62, 63]. In these works, the caryophyllene biosynthesis pathway was engineered into the carbohydrate conversion strain and the protein conversion strain respectively. The engineered coculture produced up to 507.4 mg/L of total terpene mixture including sesquiterpene, monoterpene, and caryophyllene from algae hydrolysates. Terpenes are considered potential "drop-in" candidates for aviation fuels based on their high energy density. Importantly, the terpene yield produced from the *E. coli* consortia is significantly higher than those from plant tissue.

### 4.5. Microalgae cocultures

In addition to lignocellulosic biomass, microalgae are an attractive biomass resource for biofuel production. Microalgae can provide several types of biofuels, including methane produced by anaerobic digestion of the algae biomass [64], biodiesel derived from microalgae oil [65], and biohydrogen produced photobiologically [66]. Many microalgae species can accumulate substantial quantities of lipids and contribute to a high oil yield. The average lipid content of microalgae varies between 1 and 70% and some can accumulate up to 90% of dry weight under certain conditions [65]. Furthermore, the lipids extracted from microalgae biomass often consist of triglycerides and can be converted to biodiesel by transesterification reactions in which three fatty acid molecules are esterified with a glycerol molecule. Synthetic consortia with microalgae cocultured with heterotrophs have been developed to take benefit from their mutualistic interactions.

Oleaginous yeast that can also accumulate high account of lipids has been cocultured with microalgae in several studies. In the mixed culture, yeasts use a vast variety of organic matter and provide $CO_2$ for photosynthesis by the microalgae. At the same time, microalgae act as an oxygen generator which benefits yeast growth. The mixed cultures usually showed higher yields of lipid production and higher growth rates and biomass concentration of the microalgae. Examples of oleaginous yeast and microalgae coculture include the culture of yeast *Rhodotorula glutinis* with microalgae *Chlorella vulgaris* [67] and yeast *Rhodotorula glutinis* with cyanobacterium *Spirulina platensis* [68]. In these studies, the lipid production and algae growth was reported to be higher than that in the pure cultures. Applying a similar strategy, cocultivation of filamentous fungus *A. fumigatus* and microalgae has also been demonstrated to increase biomass production and lipid yield [69]. Moreover, because filamentous fungi are bioflocculating agents, cocultivation with microalgae assisted the flocculation and therefore the harvesting of microalgae.

In the oleaginous yeast and microalgae coculture, organic carbon feedstocks are usually needed to feed the yeasts to produce biofuel products. A phototrophic sucrose-secreting cyanobacteria *Synechococcus elongatus* was engineered and cocultured with oleaginous yeast strain *Cryptococcus curvatus* or *Rhodotorula glutinis* which is capable of producing biofuel products [70]. In this case, *S. elongatus* utilized sunlight and $CO_2$ and produced sucrose as a carbon source to yeasts while yeasts assist cyanobacteria growth and survival by eliminating oxidative stress. The oleaginous yeast strain was shown to have increased lipid production in the coculture and could be engineered to produce biofuel products. This synthetic coculture presents a potential sustainable production platform for biofuels production directly from sunlight and $CO_2$.

Algae-bacterium-archaeon consortia have also been developed for the production of oil-like mixtures under anaerobic, thermophilic, and atmospheric conditions. Thermostable bacterium *Thermosipho globiformans* and archaeon *Methanocaldococcus jannaschii* were cocultured at 68°C with different species of microalgae under anaerobic conditions, followed by pyrolysis at 300°C and the consortia produced n-alkane rich biofuels and isoprenoids [71]. The composition and quantities of n-alkanes produced by pyrolysis were found to be closely related to the lipid contents and composition of the microalgae.

## 5. Tools for studying microbial consortia

Stability and tunable population compositions are highly desirable for microbial consortia developed for bioprocessing applications, because these properties could expand possible process configurations and improve efficiency. For examples, stability allows the use of continuous reactors and avoids eliminating one strain in the coculture during fermentation, whereas tunability would allow the optimization of the population composition for desired performance. Because of the multisubstrate nature of the biomass, microbial consortia with tunable population compositions are especially important to be adapted to the variable substrate compositions. Understanding the population dynamics and interactions between the members in the microbial consortia is important to develop a coculture with stable population and to tune the composition of the consortia.

Real-time PCR assays have been used to study the population dynamics of the consortia developed for biofuel production. In the study of simultaneous conversion of sugar and protein fractions of the hydrolysates by an *E. coli* coculture, specific primers targeting the unique genes in the chromosome of the two different *E. coli* strains were designed and q-PCR based quantification method was developed to monitor the temporal profile of cell growth of the two strains in the coculture during fermentation [61]. The results indicated that an optimized coculture population which was tunable by changing the inoculation ratio of the two strains is essential for the consortium to achieve higher biofuel yield. q-PCR was also used to probe the dynamics of the yeast consortium which assembled the mini-cellulosome for ethanol production [72]. The primers were designed to specifically target a unique gene encoding for the endoglucanase, exoglucanase, β-glucosidase and the scaffolding expressed by each of the yeast strain. It was found that the final population ratio of the four yeast species did not change significantly compared with the initial inoculation ratio. By comparing the population of each strain in the assembled mini-cellulosome structure with that in the free enzyme system, the synergistic effect among the cellulases in the mini-cellulosome on cellulose hydrolysis was suggested. In another study, a real-time PCR assay based on the 16S rRNA gene sequence was performed to study the population abundance of each strain in a coculture consist of a celluloytic bacterium and a noncelluloytic, solventogenic bacterium during the production of butanol [73]. The competition and cooperation relationships between the two strains at different stages of the fermentation was revealed by the population dynamics study; it was found that the population of each strain was readily modulated by culture conditions such as pH and nutrient availability.

Beyond the experimental demonstrations of bioconversion consortia, a modeling framework based on comprehensive ordinary differential equation has been developed to gain insights into the behavior and dynamics of a fungal-bacterial consortium for isobutanol production [44]. The rate expressions for each of the reaction steps were derived and the parameter values were obtained from the literature or by experiment. The concentrations of cellulose, microbial biomass, and isobutanol during the coculture fermentation predicted by the model were validated by experimental data. The model suggested that the competition between the fungal and bacterial strain for soluble saccharides is the key interaction that drives the behavior of two strains in the coculture and the relationship between the two strains was recognized as the cooperator-cheater. The model could also predict the outcomes and stability of the interactions in the microbial consortia, which provided important information for tuning the coculture population and stabilizing the consortia.

Furthermore, genome-based metabolic networks using multispecies dynamic flux balance analysis were developed to build a process model for an open pond system involving the oleaginous yeast and microalgae consortia for biodiesel production [74]. The algal monoculture and yeast monoculture were modeled separately and compared to the algae and yeast coculture with cellulosic glucose and xylose feeds. The model predicted the biomass and lipid productivities of the coculture with results comparable to those reported in literature. The economic analysis of this system was also performed and indicated that the algae and yeast coculture can produce biodiesel at competitive prices.

## 6. Conclusion

The development and implementation of biorefineries using biomass as the raw materials is encouraged by the growth in demand for renewable fuels and chemicals and the need for a reduction of fossil energy derived green-house gas emissions. Challenges facing scale-up of the general biorefinery concept include technoeconomic hurdles stemming from recalcitrance of the biomass to hydrolysis, feedstock variability, achieving high rate, yield, and titers of bioconversion of the bulk of the biomass to fuels, up- and down-stream separations, and minimizing resource inputs, including water, chemical additives, electricity, and infrastructure. Progress in development of consortium-based bioconversion technologies provides solutions to many of these challenges by consolidating pretreatment and biocatalysis, allowing flexibility for utilization of multiple substrates at variable concentrations, and supporting tunability for targeted end-products. New advances in synthetic biology and metabolic engineering in the context of microbial communities will be required to accelerate adoption and scale-up of these strategies for an economically viable bio-based economy.

## Acknowledgements

Sandia National Laboratories is a multimission laboratory managed and operated by National Technology and Engineering Solutions of Sandia LLC., a wholly owned subsidiary of Honeywell International Inc. for the US Department of Energy's National Nuclear Security Administration under contract DE-NA0003525.

## Conflict of interest

The authors declare that they have no competing interests.

## Author details

Fang Liu*, Eric Monroe and Ryan W. Davis

*Address all correspondence to: fliu006@gmail.com and fanliu@sandia.gov

Department of Biomass Science & Conversion Technologies, Sandia National Laboratories, Livermore, CA, USA

## References

[1] Langholtz MH, Stokes BJ, Eaton LM. 2016 Billion-Ton Report: Advancing domestic resources for a thriving bioeconomy, Volume 1: Economic availability of feedstocks. Oak Ridge National Laboratory, Oak Ridge, TN; 2016, 448p. DOI: 10.2172/1271651

[2] Schwab A. Bioenergy Technologies Office Multi-Year Program Plan. March 2016. United States; 2016. DOI:10.2172/1245338

[3] McCann MC, Carpita NC. Biomass recalcitrance: A multi-scale, multi-factor, and conversion-specific propertyo. Journal of Experimental Botany. 2015;**66**:4109-4118

[4] Kumar R, Singh S, Singh OV. Bioconversion of lignocellulosic biomass: Biochemical and molecular perspectives. Journal of Industrial Microbiology and Biotechnology. 2008;**35**:377-391

[5] Eiteman MA, Lee SA, Altman E. A co-fermentation strategy to consume sugar mixtures effectively. Journal of Biological Engineering. 2008;**2**:1-8

[6] Wu W, Liu F, Singh S. Toward engineering *E. coli* with an autoregulatory system for lignin valorization. Proceedings of the National Academy of Sciences. 2018;**115**:2970-2975

[7] Cherubini F. The biorefinery concept: Using biomass instead of oil for producing energy and chemicals. Energy Conversion and Management. 2010;**51**:1412-1421

[8] Athaley SSA, Zheng W, Saha MIDB. Process intensification for cellulosic biorefineries. ChemSusChem. 2017;**10**:2566-2572

[9] Morris BE, Henneberger R, Huber H, Moissl-Eichinger C. Microbial syntrophy: Interaction for the common good. FEMS Microbiology Reviews. 2013;**37**:384-406

[10] Seth EC, Taga ME. Nutrient cross-feeding in the microbial world. Frontiers in Microbiology. 2014;**5**:1-6

[11] Moon RJ, Martini A, Nairn J, Simonsen J, Youngblood J. Cellulose nanomaterials review: Structure, properties and nanocomposites. Chemical Society Reviews. 2011;**40**:3941-3994

[12] Puls J. Chemistry and biochemistry of hemicelluloses: Relationship between hemicellulose structure and enzymes required for hydrolysis. Macromolecular Symposia. 1997;**120**:183-196

[13] Welker CM, Balasubramanian VK, Petti C, Rai KM, De Bolt S, Mendu V. Engineering plant biomass lignin content and composition for biofuels and bioproducts. Energies. 2015;**8**:7654-7676

[14] Eudes A, Liang Y, Mitra P, Loqué D. Lignin bioengineering. Current Opinion in Biotechnology. 2014;**26**:189-198

[15] Kumar P, Barrett DM, Delwiche MJ, Stroeve P. Methods for pretreatment of lignocellulosic biomass for efficient hydrolysis and biofuel production. Industrial and Engineering Chemistry. 2009;**48**:3713-3729

[16] Hannon M, Gimpel J, Tran M, Rasala B, Mayfield S. Biofuels from algae: Challenges and potential. Biofuels. 2010;**1**:763-784

[17] Chisti Y. Biodiesel from microalgae beats bioethanol. Trends in Biotechnology. 2008;**26**: 126-131

[18] Greenwell HC, Laurens LML, Shields RJ, Lovitt RW, Flynn KJ. Placing microalgae on the biofuels priority list: A review of the technological challenges. Journal of the Royal Society Interface. 2010;**7**:703-726

[19] Brown MR, Jeffrey SW, Volkman JK, Dunstan GA. Nutritional properties of microaglae for mariculture. Aquaculture. 1997;**151**:315-331

[20] Markou G, Angelidaki I, Georgakakis D. Microalgal carbohydrates: An overview of the factors influencing carbohydrates production, and of main bioconversion technologies for production of biofuels. Applied Microbiology and Biotechnology. 2012;**96**:631-645

[21] Capelli B, Cysewski GR. Potential health benefits of spirulina microalgae. Nutrafoods. 2010;**9**:19-26

[22] Kenney KL, Smith WA, Gresham GL, Westover TL. Understanding biomass feedstock variability. Biofuels. 2013;**4**:111-127

[23] Mourtzinis S, Cantrell KB, Arriaga FJ, Balkcom KS, Novak JM, Frederick JR, et al. Carbohydrate and nutrient composition of corn stover from three southeastern USA locations. Biomass and Bioenergy. 2016;**85**:153-158

[24] Liu K. Chemical composition of distillers grains, a review. Journal of Agricultural and Food Chemistry. 2011;**59**:1508-1526

[25] Belyea RL, Rausch KD, Tumbleson ME. Composition of corn and distillers dried grains with solubles from dry grind ethanol processing. Bioresource Technology. 2004;**94**: 293-298

[26] Liu K. Particle size distribution of distillers dried grains with solubles (DDGS) and relationships to compositional and color properties. Bioresource Technology. 2008;**99**: 8421-8428

[27] Cromwell GL, Herkelman KLST. Physical, chemical, and nutritional characteristics of distillers dried grains with solubles for chicks and pigs. Journal of Animal Science. 1993;**71**:679-686

[28] Spiehs MJ, Whitney MHSG. Nutrient database for distiller's dried grains with solubles produced from new ethanol plants in Minnesota and South Dakota. Journal of Animal Science. 2002;**80**:2639-2645

[29] Davis R, Markham J, Kinchin C, Grundl N, Tan ECD, Humbird D. Process design and economics for the production of algal biomass: Algal biomass production in open pond systems and processing through dewatering for downstream conversion. National Renewable Energy Laboratory, NREL/TP-5100-64772. 2016

[30] Chubukov V, Mukhopadhyay A, Petzold C, Keasling J. Synthetic and systems biology for microbial production of commodity chemicals : From target selection to scale-up. NPJ Systems Biology and Applications. 2016;**16009**:1-11

[31] Liao JC, Mi L, Pontrelli S, Luo S. Fuelling the future: Microbial engineering for the production of sustainable biofuels. Nature Reviews. Microbiology. 2016;**14**:288-304

[32] Kim H, Sun Q, Liu F, Tsai SL, Chen W. Biologically assembled nanobiocatalysts. Topics in Catalysis. 2012;**55**:1138-1145

[33] Liu F, Banta S, Chen W. Functional assembly of a multi-enzyme methanol oxidation cascade on a surface-displayed trifunctional scaffold for enhanced NADH production. Chemical Communications. 2013;**49**:3766

[34] Park M, Sun Q, Liu F, DeLisa MP, Chen W. Positional assembly of enzymes on bacterial outer membrane vesicles for cascade reactions. PLoS One. 2014;**9**:1-6

[35] Bernstein HC, Carlson RP. Microbial consortia engineering for cellular factories: In vitro to in silico systems. Computational and Structural Biotechnology Journal. 2012;**3**: e201210017

[36] Jia X, Liu C, Song H, Ding M, Du J, Ma Q, Yuan Y. Design, analysis and application of synthetic microbial consortia. Synthetic and Systems Biotechnology. 2016;**1**:109-117

[37] Brenner K, You L, Arnold FH. Engineering microbial consortia: A new frontier in synthetic biology. Trends in Biotechnology. 2008;**26**:483-489

[38] Pappenhagen J. Biological Treatment of Sewage and Industrial Wastes. New York: Reinhold Publishing; 1957

[39] Conrad R. Contribution of hydrogen to methane production and control of hydrogen cocnentrations in methanogenic soils. FEMS Microbial Ecology. 1999;**28**:193-202

[40] Haruta S, Nakayama T, Nakamura K, Hemmi H, Ishii M, Igarashi Y, et al. Microbial diversity in biodegradation and reutilization processes of garbage. Journal of Bioscience and Bioengineering. 2005;**99**:1-11

[41] Ward AJ, Lewis DM, Green FB. Anaerobic digestion of algae biomass: A review. Algal Research. 2014;**5**:204-214

[42] Bayer EA, Belaich J-P, Shoham Y, Lamed R. The cellulosomes: Multienzyme machines for degradation of plant cell wall polysaccharides. Annual Review of Microbiology. 2004;**58**:521-554

[43] Argyros DA, Tripathi SA, Barrett TF, Rogers SR, Feinberg LF, Olson DG, et al. High ethanol titers from cellulose by using metabolically engineered thermophilic, anaerobic microbes. Applied and Environmental Microbiology. 2011;**77**:8288-8294

[44] Minty JJ, Singer ME, Scholz SA, Bae C-H, Ahn J-H, Foster CE, et al. Design and characterization of synthetic fungal-bacterial consortia for direct production of isobutanol from cellulosic biomass. Proceedings of the National Academy of Sciences. 2013;**110**:14592-14597

[45] Bokinsky G, Peralta-Yahya PP, George A, Holmes BM, Steen EJ, Dietrich J, et al. Synthesis of three advanced biofuels from ionic liquid-pretreated switchgrass using engineered *Escherichia coli*. Proceedings of the National Academy of Sciences. 2011;**108**:19949-19954

[46] Bayer TS, Widmaier DM, Temme K, Mirsky E, Santi DV, Voigt C. Synthesis of methyl halides from biomass using engineered microbes synthesis of methyl halides from biomass using engineered microbes. Journal of the American Chemical Society. 2009; **131**:6508-6515

[47] Tsai SL, Goyal G, Chen W. Surface display of a functional minicellulosome by intracellular complementation using a synthetic yeast consortium and its application to cellulose hydrolysis and ethanol production. Applied and Environmental Microbiology. 2010;**76**:7514-7520

[48] Masran R, Zanirun Z, Bahrin EK, Ibrahim MF, Lai Yee P, Abd-Aziz S. Harnessing the potential of ligninolytic enzymes for lignocellulosic biomass pretreatment. Applied Microbiology and Biotechnology. 2016;**100**:5231-5246

[49] Stulke J, Hillen W. Carbon catabolite repression in bacteria. Current Opinion in Microbiology. 1999;**2**:195-201

[50] Ho NWY, Chen Z, Brainard AP. Genetically engineered Saccharomyces yeast capable of effective cofermentation of glucose and xylose. Applied and Environmental Microbiology. 1998;**64**:1852-1859

[51] Nichols NN, Dien BS, Bothast RJ. Use of catabolite repression mutants for fermentation of sugar mixtures to ethanol. Applied Microbiology and Biotechnology. 2001;**56**:120-125

[52] Kuyper M, Toirkens MJ, Diderich JA, Winkler AA, Van Dijken JP, Pronk JT. Evolutionary engineering of mixed-sugar utilization by a xylose-fermenting *Saccharomyces cerevisiae* strain. FEMS Yeast Research. 2005;**5**:925-934

[53] Eiteman MA, Lee SA, Altman R, Altman E. A substrate-selective co-fermentation strategy with *Escherichia coli* produces lactate by simultaneously consuming xylose and glucose. Biotechnology and Bioengineering. 2009;**102**:822-827

[54] Xia T, Eiteman MA, Altman E. Simultaneous utilization of glucose, xylose and arabinose in the presence of acetate by a consortium of *Escherichia coli* strains. Microbial Cell Factories. 2012;**11**:1

[55] Saini M, Lin LJ, Chiang CJ, Chao YP. Synthetic consortium of *Escherichia coli* for n-butanol production by fermentation of the glucose-xylose mixture. Journal of Agricultural and Food Chemistry. 2017;**65**:10040-10047

[56] Caballero A, Ramos JL. Enhancing ethanol yields through D-xylose and L-arabinose co-fermentation after construction of a novel high efficient L-arabinose-fermenting Saccharomyces cerevisiae strain. Microbiology. 2017;**163**:442-452

[57] Zhang H, Pereira B, Li Z, Stephanopoulos G. Engineering *Escherichia coli* coculture systems for the production of biochemical products. Proceedings of the National Academy of Sciences. 2015;**112**:8266-8271

[58] Bothast RJ, Schlicher MA. Biotechnological processes for conversion of corn into ethanol. Applied Microbiology and Biotechnology. 2005;**67**:19-25

[59] Huo Y-X, Cho KM, Rivera JGL, Monte E, Shen CR, Yan Y, et al. Conversion of proteins into biofuels by engineering nitrogen flux. Nature Biotechnology. 2011;**29**:346-351

[60] Wu W, Tran-Gyamfi MB, Jaryenneh JD, Davis RW. Cofactor engineering of ketol-acid reductoisomerase (IlvC) and alcohol dehydrogenase (YqhD) improves the fusel alcohol yield in algal protein anaerobic fermentation. Algal Research. 2016;**19**:162-167

[61] Liu F, Wu W, Tran-Gyamfi MB, Jaryenneh JD, Zhuang X, Davis RW. Bioconversion of distillers' grains hydrolysates to advanced biofuels by an *Escherichia coli* co-culture. Microbial Cell Factories. 2017;**16**:192

[62] Wu W, Liu F, Davis RW. Engineering *Escherichia coli* for the production of terpene mixture enriched in caryophyllene and caryophyllene alcohol as potential aviation fuel compounds. Metabolic Engineering Communications. 2018;**6**:13-21

[63] Wu W, Davis RW. One-pot bioconversion of algae biomass into terpenes for advanced biofuels and bioproducts. Algal Research. 2016;**17**:316-320

[64] Hernández D, Solana M, Riaño B, García-González MC, Bertucco A. Biofuels from microalgae: Lipid extraction and methane production from the residual biomass in a biorefinery approach. Bioresource Technology. 2014;**170**:370-378

[65] Martins A, Caetano NS, Mata TM. Microalgae for biodiesel production and other applications: A review. Renewable and Sustainable Energy Reviews. 2010;**14**:217-232

[66] Ghirardi ML, Zhang L, Lee JW, Flynn T, Seibert M, Greenbaum E, et al. Microalgae: A green source of renewable $H_2$. Trends in Biotechnology. 2000;**18**:506-511

[67] Cheirsilp B, Suwannarat W, Niyomdecha R. Mixed culture of oleaginous yeast Rhodotorula glutinis and microalga *Chlorella vulgaris* for lipid production from industrial wastes and its use as biodiesel feedstock. New Biotechnology. 2011;**28**:362-368

[68] Xue F, Miao J, Zhang X, Tan T. A new strategy for lipid production by mix cultivation of *Spirulina platensis* and *Rhodotorula glutinis*. Applied Biochemistry and Biotechnology. 2010;**160**:498-503

[69] Wrede D, Taha M, Miranda AF, Kadali K, Stevenson T, Ball AS, et al. Co-cultivation of fungal and microalgal cells as an efficient system for harvesting microalgal cells, lipid production and wastewater treatment. PLoS One. 2014;**9**:1-22

[70] Li T, Li CT, Butler K, Hays SG, Guarnieri MT, Oyler GA, et al. Mimicking lichens: Incorporation of yeast strains together with sucrose-secreting cyanobacteria improves survival, growth, ROS removal, and lipid production in a stable mutualistic co-culture production platform. Biotechnology for Biofuels. 2017;**10**:1-11

[71] Yamane K, Matsuyama S, Igarashi K, Utsumi M, Shiraiwa Y, Kuwabara T. Anaerobic coculture of microalgae with thermosipho globiformans and methanocaldococcus jannaschii at 68°C enhances generation of n-alkane-rich biofuels after pyrolysis. Applied and Environmental Microbiology. 2013;**79**:924-930

[72] Goyal G, Tsai SL, Madan B, DaSilva NA, Chen W. Simultaneous cell growth and ethanol production from cellulose by an engineered yeast consortium displaying a functional mini-cellulosome. Microbial Cell Factories. 2011;**10**:1-8

[73] Wen Z, Wu M, Lin Y, Yang L, Lin J, Cen P. Artificial symbiosis for acetone-butanol-ethanol (ABE) fermentation from alkali extracted deshelled corn cobs by co-culture of *Clostridium beijerinckii* and *Clostridium cellulovorans*. Microbial Cell Factories. 2014;**13**:1-11

[74] Gomez JA, Höffner K, Barton PI. From sugars to biodiesel using microalgae and yeast. Green Chemistry. 2016;**18**:461-475

Section 3

# Impacts

Chapter 8

# Effect of Binary Fuel Blends on Compression Ignition Engine Characteristics: A Review

Paramvir Singh, S.R. Chauhan and Varun Goel

Additional information is available at the end of the chapter

http://dx.doi.org/10.5772/intechopen.80566

**Abstract**

Diversified research in alternate sources arises become necessity due to higher consumption of fossil fuels along with their adverse impacts on the environment, even to the point of complete elimination of diesel from compression-ignition (CI) engines. Binary fuel blend (a blend of low and high viscous fuel) is one of the best environmentally friendly alternative in CI engines. Blending of methyl ester with edible and nonedible oils in different volumetric ratios has the potency to give a stable mixture and that can be used as a fuel in diesel engines. The main motive for the blending of two fuels is that the inferior properties of one biofuel remunerate from improved properties of the other fuel considerably improves the physicochemical properties of the blend. The present study provides comprehensive information on the emission and performance characteristics of binary biodiesel-oil fuel blends. Most researchers had suggested optimum blends from their respective studies that support capability for complete elimination of diesel from CI engines. Some researchers have used this binary fuel blend with minor adjustments to the engine parameters. These investigations have provided positive results. The comprehensive review concluded that binary fuel approach has potential to completely eliminate diesel from CI engines.

**Keywords:** binary fuel blends, emission profile, fossil fuel depletion, performance analysis

## 1. Introduction

Demand of energy has been increased day by day due to the usage of nearly 600 million automobiles by 7 billion people throughout the world [1]. Today, the air contamination is also a severe concern as higher pollutants in today's environment are from internal combustion (IC)

engine tailpipe emissions. Hence, the automobile industries are emphasizing on cleaner transport systems that decrease the environmental pollution and provide at least the same efficiency as normal diesel fuel [2]. Also, the pressure of environmental monitoring authorities forces the automobile companies to diminish the greenhouse gas emissions from the vehicles. These ever-increasing energy demand, rapid depletion of crude oil, and environmental pollution have an emphasis on a new feasible source of energy known as biofuel (mainly biodiesel). The another reason for emphasis on alternate fuel which can be produced from readily accessible resources is that the countries which do not have sufficient resources of fossil fuel and import crude petroleum are facing foreign exchange problem or energy crisis. Therefore, the exploration is focused on numerous feedstocks (edible/nonedible oils along with its methyl esters (biodiesel)) as a probable source of fuel for automobile sector [3]. It is better to use methyl ester instead of fossil-based diesel fuel due to its various benefits, that is, availability, high lubricity, low sulfur contents, low aromatic compounds, environmentally friendly nature, the existence of a carbon dioxide ($CO_2$) cycle in burning period, viability, and biodegradability [4–9]. In general, biodiesel is clean burning mono-alkyl ester-based oxygenated fuel derived from long chain fatty acids of edible and nonedible oil, animal fat, microalgae, etc. [10]. The main application of biodiesel is in diesel engines as a fuel had shown that it lowers GHG emissions with the acceptable engine performance [11]. This finally leads us to the study of diesel engine characteristics of biodiesels, since they are the single most important contender in the area of future fuels.

Broad research has been carried out with the aim of using biodiesel and its blends as a feedstock for diesel engine. In this concern, B20 (80% diesel and 20% biodiesel by volume) is found to be the most suitable blend without any modification to the engine hardware [12]. Some researchers had used biodiesel-diesel-alcohol blends for increasing the biodiesel contents in the blend and conveyed that blending of 25% biodiesel-5% alcohol and 70% diesel give better result in terms of emission and performance in diesel engines [13]. However, with the application of 20 and 25% biodegradable realm in diesel engine, it is not presumable to attain the fossil fuel nondependence energy sources. Furthermore, the sustainability as well as the environmental friendliness of these blends decreases as the percentage of fossil-based diesel fuel is more in the blend as compared to diesel. So, it is imperative to take a step toward the total replacement of diesel from CI engines. In this regard, some work has been done on biodiesel-alcohol, biodiesel-oil, and biodiesel-biodiesel blends. The different approaches of the researchers who are working in the area of alternate fuels for engine application with the aim of partial and complete elimination of diesel is tabulated in **Table 1**. The applicability of neat biodiesel in diesel engine is also too low as it affects the engine performance and its running life due to high viscosity. One of the best possible ways to achieve the above stated requirement is to use binary fuel (biodiesel-oil) blends which possess properties as per American Society for Testing Materials (ASTM) specification for biodiesel. Using methyl ester (high viscous) in various proportions with nonedible oils (low viscous) have the ability to give a stable solution and feasible to completely replace diesel from CI engines [1]. The biodiesel can be a solubilizer and improve properties of the blend [14]. The major advantage of this binary fuel blend is that it can be used in CI engines without any major tweaking, and second, it

| | | |
|---|---|---|
| Elimination of conventional fuel from diesel engines | Partial elimination | Biodiesel + diesel blends |
| | | Biodiesel + alcohol + diesel blends |
| | | Biodiesel 1 + biodiesel 2 + diesel blends |
| | Complete elimination | Biodiesel + alcohol blends |
| | | Neat biodiesel |
| | | Biodiesel + oil blends |
| | | Other mixed feedstock |

**Table 1.** Overview of elimination of conventional fuel from CI engines.

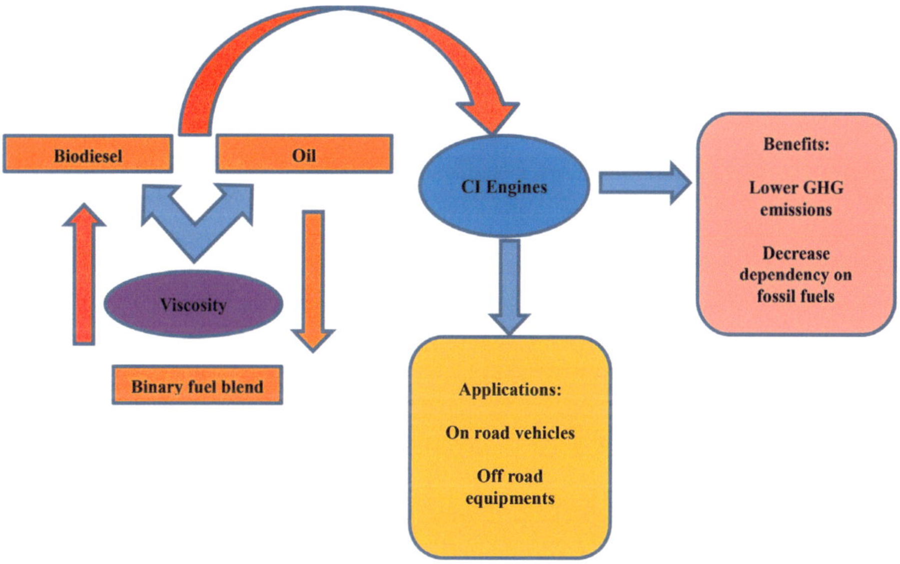


**Figure 1.** Graphical representation of binary fuel approach.

produces less harmful gases as emission to the environment. Another advantage of using binary biofuel is that it also decreases our dependency on fossil fuels without compromising with the efficiency of the engine. Also, the feedstock used in binary fuel approach is renewable in nature. The graphical representation of binary fuel approach is shown in **Figure 1**.

The present study is to highlight the prospective of biodiesel with oil blend to be used as a feedstock for diesel engine in the automotive sector and can also make available effectively for the complete elimination of diesel from standing engine with these blends. This contributes to the constraints made by organizations like Occupational Safety and Health Administration (OSHA) and Environmental Protection Agency (EPA) toward the replacement of fossil fuel and finding an alternative that is environmental friendly and biodegradable.

## 2. Performance analysis and emission profile of binary fuel blends

Blending of biodiesel with low viscous oil has a potential to completely replace diesel from the CI engine, and some work was already done on this type of blending.

Devan and Mahalakshmi [14] had used paradise oil methyl ester (PME) blends with eucalyptus (EU) oil in the proportion of 20, 30, 40 and 50% on a volume basis. Outcome revealed that 50% blend of PME and 50% EU is an optimum blend. The authors found that this blend shows a significant reduction in unburnt hydrocarbons (HC), carbon monoxide (CO), and smoke opacity at all load conditions. But, there is some increase in nitrogen oxide ($NO_x$) observed. Although the enhancement in brake thermal efficiency (BTE), 2.4% at full load condition and decrement in most of the tailpipe emissions overshadow the slight increase in $NO_x$. Vallinayagam et al. [15] had also assessed diesel engine characteristics fuelled with biodiesel oil blends. They had used blends of kapok methyl ester (KME), that is, high viscous and pine oil (low viscous) in the proportion of 25, 50, and 75% on the volume basis in single cylinder 4-stroke water cooled CI engine. Results revealed that 50% KME and 50% pine oil blend on volume basis (KME50P50) shows optimum result for engine performance and emission. With this blend as a fuel in diesel engine, they found 12.5, 18.9, and 8.1% decrement in CO, smoke, and HC emissions as compared to diesel, respectively. Results revealed that exhaust gas temperature (EGT) also reduced minimally at all the load conditions. And $NO_x$ emissions were found in line with diesel. They also found the performance of KME50P50 similar to conventional diesel fuel at high load condition perhaps its slight lower than diesel at low load conditions. Sharma and Murugan [16] had investigated Jatropha oil biodiesel with tyre pyrolysis oil (TPO) in composition as a fuel. They were found out that 80% Jatropha biodiesel-20% tyre pyrolysis oil blend provides better result in terms of performance and emissions and stated it as an optimum blend. Dubey and Gupta [17] had also used Jatropha methyl ester. They investigated diesel engine characteristics with Jatropha methyl ester blend and turpentine oil blends, that is, JB90TO10, JB70TO30, and JB50TO50. JB50TO50 shows improved results as compared to other tested blends. At full load condition, tailpipe emissions, that is, HC, CO, smoke, and $NO_x$ were found to be 42.5, 4.56, 29.16, and 4.72%, respectively, lower than conventional diesel although $CO_2$ emissions rises by 10.5%. Singh et al. [1] had used Amla seed oil biodiesel (AB) and EU in the various proportions of 9:1, 8:2, 7:3, 6:4, and 5:5 by volume. They concluded that among all the tested blends, AB70EU30 is optimum blend as it gives better results in terms of combustion, performance, and emission characteristics. AB70EU30 shows better results in terms of emissions and comparable performance as diesel at high load conditions.

All the authors used biodiesel-oil blends as a feedstock for diesel engine. They maximally found that the emissions decrease drastically with the use of biodiesel-oil blends, and one of the optimum blends from the tested blends has better combustion and emission characteristics. The effect on performance and emission characteristics of diesel engine with different binary fuel blends is due to variable properties of the fuels. The properties of the fuel used and their binary fuel blends are given in **Table 2**. The effect of operating conditions on performance and emission characteristics of different binary fuel blends is shown in **Table 3**, and the optimum blends suggested by the authors is tabulated in **Table 4**.

| Properties / Fuels | Density @40°C in kg/m$^3$ | Kinematic viscosity @ 40°C in CST | Conradson carbon residue | Fire point (°C) | Flash point (°C) | Pour point (°C) | Heating value (kJ/kg) | Sulfur wt% | Saponification value | Iodine value | Distillation recovery @90% min | Cetane number | Ref. |
|---|---|---|---|---|---|---|---|---|---|---|---|---|---|
| IS 1460–1974 | Nil | 2.0–7.5 | 0.20 | — | 38 min | 6 max | — | 1.00 max | — | — | 366°C | 42 | [14] |
| Diesel | 0.84 | 3–4 | — | — | 74 | −23 | 42,700 | — | — | — | — | 40–55 | |
| PME | 0.8752 | 5.4 | 0.18 | 150 | 141.2 | 2 | 40,285 | 0.13 | 191.5 | 46 | 369 | 51 | |
| EU | 0.8955 | 2.0 | 1.90 | — | 54 | −5 | 43,270 | — | — | — | — | — | |
| PME20EU80 | 0.8914 | 2.72 | — | — | — | — | 42,673 | — | — | — | — | — | |
| PME30EU70 | 0.8894 | 3.08 | — | — | — | — | 42,374 | — | — | — | — | — | |
| PME40EU60 | 0.8874 | 3.44 | — | — | — | — | 42,076 | — | — | — | — | — | |
| PME50EU50 | 0.8852 | 3.8 | — | — | — | — | 41,778 | — | — | — | — | — | |
| KME | 0.875 | 5.4 | | — | 156 | — | 36,292 | <0.005 | — | — | — | 54 | [15] |
| Pine oil | 0.8751 | 1.3 | — | — | 52 | — | 42,800 | <0.005 | — | — | — | 11 | |
| KME25P75 | 0.875 | 2.3 | — | — | 78 | — | 41,173 | <0.005 | — | — | — | 22 | |
| KME50P50 | 0.875 | 3.3 | — | — | 104 | — | 39,546 | <0.005 | — | — | — | 33 | |
| KME75P25 | 0.875 | 4.4 | — | — | 130 | — | 37,920 | <0.005 | — | — | — | 43 | |
| JB | 0.880 | 5.65 | 0.5 | | 170 | — | 38,450 | — | — | — | — | 50–55 | [16] |
| TPO | 0.920 | 5.4 | — | 50 | 43 | — | 39,200 | — | — | — | — | — | |
| JB90TPO10 | 0.883 | 5.73 | — | 78 | 64 | — | 39,240 | — | — | — | — | — | |
| JB80TPO20 | 0.887 | 5.60 | — | 73 | 60 | — | 37,740 | — | — | — | — | — | |
| JB70TPO30 | 0.892 | 5.41 | — | 69 | 55 | — | 36,400 | — | — | — | — | — | |
| JB60TPO40 | 0.894 | 5.29 | — | 64 | 49 | — | 35,120 | — | — | — | — | — | |
| JB50TPO50 | 0.897 | 5.15 | — | 59 | 44 | — | 33,890 | — | — | — | — | — | |
| AB | 0.884 | — | 0.04 | — | 152 | −0.5 | 40,100 | — | — | — | — | — | [1] |
| TO | 0.920 | 4.12 | — | — | 38 | −23 | 44,400 | — | — | — | — | 38 | [17] |
| JB50TO50 | 0.900 | 4.01 | — | — | — | — | 41,950 | — | — | — | — | — | |
| JB70TO30 | 0.892 | 4.08 | — | — | — | — | 40,480 | — | — | — | — | — | |
| JB90TO10 | 0.885 | 4.12 | — | — | — | — | 39,990 | — | — | — | — | — | |

**Table 2.** Properties of the fuels and binary fuel blends.

| Fuel blends used | Base fuel | Engine | Operating conditions | Performance | | | Emissions | | | | | | Ref. |
|---|---|---|---|---|---|---|---|---|---|---|---|---|---|
| | | | | BSEC | BSFC | BTE | HC | $NO_X$ | CO | $CO_2$ | Smoke | EGT | |
| PME20EU80 | Diesel | 1 cylinder 4-stroke air cooled diesel engine | At a constant speed of 1500 rpm and different loads | ⇩ | – | ⇨/⇧ | ⇩ | ⇧ | ⇨/⇩ | – | ⇩ | ⇧ | [14] |
| PME30EU70 | | | | ⇧ | – | ⇨/⇩ | ⇩ | ⇧ | ⇧/⇩ | – | ⇩ | ⇧ | |
| PME40EU60 | | | | ⇧ | – | ⇨/⇩ | ⇩ | ⇧ | ⇧/⇩ | – | ⇩ | ⇧ | |
| PME50EU50 | | | | ⇩ | – | ⇨/⇧ | ⇩ | ⇧ | ⇧/⇩ | – | ⇩ | ⇧ | |
| KME100 | Diesel | Single cylinder, 4 stroke, DI diesel engines | Constant speed of 1500 rpm and different loads | – | ⇧ | ⇩ | – | ⇧ | – | – | ⇧ | ⇧ | [15] |
| KME25P75 | | | | – | ⇩ | ⇧ | – | ⇩/⇧ | ⇧/⇩ | – | ⇩ | ⇩ | |
| KME50P50 | | | | – | ⇧ | ⇧ | – | ⇩/⇧ | ⇧/⇩ | – | ⇨/⇧ | ⇩ | |
| KME75P25 | | | | – | ⇧ | ⇩ | – | ⇩ | ⇧ | – | ⇧ | ⇧ | |
| JB100 | Diesel | Single cylinder, 4 stroke, air cooled DI diesel engines | Constant speed of 1500 rpm and different loads | ⇧ | – | ⇩/⇧/⇩ | ⇩ | ⇧ | ⇩ | – | ⇧ | ⇧ | [16] |
| JB90TPO10 | | | | ⇧ | – | ⇩ | ⇩ | ⇧ | ⇩ | – | ⇩ | ⇧/⇩ | |

| Fuel blends used | Base fuel | Engine | Operating conditions | Performance | | | Emissions | | | | | | Ref. |
|---|---|---|---|---|---|---|---|---|---|---|---|---|---|
| | | | | BSEC | BSFC | BTE | HC | $NO_X$ | CO | $CO_2$ | Smoke | EGT | |
| JB80TPO20 | | | | ↑ | – | →/↑/→ | ↓/↑/↓ | ↑ | ↓ | – | ↓ | ↑ | |
| JB70TPO30 | | | | ↑ | – | ↑/↓ | ↑ | ↑/↓/↑ | ↑ | – | ↑/↓ | ↑ | |
| JB60TPO40 | | | | ↑ | – | ↑/↓ | ↑ | ↑/↓/↑ | ↑ | – | ↑ | ↑ | |
| JB50TPO50 | | | | ↑ | – | ↑/↓ | ↑ | ↑/↓/↑ | ↑ | – | ↑ | ↑ | |
| JB100 | Diesel | Single cylinder, DI, water cooled, naturally aspirated engine | Constant speed and different load conditions (no load, 35%, 65% and full load) | – | ↑ | ↓ | ↓ | ↑ | ↓ | ↑ | ↓ | – | [17] |
| JB50T50 | | | | – | ↑ | ↓ | ↓ | ↓ | ↓ | ↑ | ↓ | – | |
| JB70T30 | | | | – | ↑ | ↓ | ↓ | ↑ | ↓ | ↑ | ↓ | – | |
| JB90T10 | | | | – | ↑ | ↓ | ↓ | ↑ | ↓ | ↑ | ↓ | – | |
| AB100 | Diesel | Single cylinder, DI, water cooled, naturally aspirated engine | Constant speed and variable load | – | ↑ | ↓ | ↑ | ↓ | ↓ | – | ↓ | ↓ | [1] |
| AB90EU10 | | | | – | ↑ | ↓ | ↑ | ↓ | ↓ | – | ↓ | ↓ | |
| AB80EU20 | | | | – | ↑ | ↓ | ↓ | ↓ | ↓ | – | ↓ | ↓ | |
| AB70EU30 | | | | – | → | ↑ | ↓ | ↓ | ↓ | – | ↓ | ↓ | |

| Fuel blends used | Base fuel | Engine | Operating conditions | Performance | | | Emissions | | | | | | Ref. |
|---|---|---|---|---|---|---|---|---|---|---|---|---|---|
| | | | | BSEC | BSFC | BTE | HC | $NO_X$ | CO | $CO_2$ | Smoke | EGT | |
| AB60EU40 | | | | – | ⇩ | ⇧ | ⇩ | ⇧ | ⇩ | – | ⇩ | ⇧ | |
| AB50EU50 | | | | – | ⇩ | ⇧ | ⇩ | ⇧ | ⇩ | – | ⇩ | ⇧ | |

First arrow/second arrow shows the change at low load and high loads.
First arrow/second arrow/third arrow shows the change at low, medium and high loads.
Single arrow shows the overall effect from lower to higher loads.

**Table 3.** Performance and emission characteristics of dual biofuels.

| Authors (Ref.) | Devan and Mahalakshmi [14] | Vallinaygam et al. [15] | Sharma and Murugan [16] | Dubey and Gupta [17] | Singh et al. [1] |
|---|---|---|---|---|---|
| Optimum blend | PME50EU50 | KME50P50 | JB80TPO20 | JB50TO50 | AB70EU30 |

**Table 4.** Optimum blend that have capability to completely eliminate diesel from CI engines.

## 3. Effect of minor modification on performance of optimum binary fuel blend

The most important parameter is the air-fuel mixing that affects the performance of biodiesel and its blends. A way to improve the air-fuel mixing is to alter the engine parameters. In this regard, some researchers have used this binary fuel blend with minor adjustments to the engine parameters.

There is very less literature available with biodiesel-oil blend as a feedstock in a diesel engine with minor tweaking. Authors [18–21] had performed experiments to study the behavior of a diesel engine running on the optimum binary fuel blends at varying compression ratio (CR), injection timing (IT), and injection pressure (IP).

CR has been altered without changing the geometry of combustion chamber with the provision of arrangement provided in variable compression engine [21]. In a simple diesel engine, CR of the engine has been changed by altering the clearance volume, by the replacement of gaskets of variable thickness in between the cylinder and the cylinder head. The fuel injection strategy is an important parameter in diesel engines to optimize the combustion, performance, and tailpipe emissions. Three IP values were used for experimentation, that is, 180, 210, and 240 bar. IP was adjusted by regulating spring tension of needle provided in injector. Injection timing was adjusted on three different values of 20, 23, and 26° bTDC by introducing required shims at the position between the fuel pump and engine. To advance the fuel IT, the shims under the pump were removed, and to retard additional shims were introduced under fuel injection pump. At standard IT, the number of shims placed under the pump was three. The thickness of the each shim is 0.3 mm and removing one shim advanced IT about 1.5°CA IT and introducing a shim retard the timing by 1.5°CA.

CR is the most valuable factors for diesel engine operation because of its high anti-knocking property. Sharma and Murugan [18] had altered the CR and compared the results of optimum Jatropha biodiesel-tyre pyrolysis oil blend. They operated the engine with diesel as a fuel at original condition and JB50TPO50 blend at altered CR conditions. They found that BTE improves unrelatedly of the engine load at higher CR. The consumption of energy at higher CR reduces at higher CR as compared to the original conditions. This may be due to the reason of better fuel spray characteristics and improved air-fuel mixing at higher CR. As compared to original CR conditions, the tailpipe emissions, that is, CO and HC to the environment from diesel engine exhaust fueled with biodiesel oil blends reduces. Smoke opacity also reduces as compared to original conditions. This is also due to the fact that air move in during the suction stroke at higher CR is compressed, which escalates the air temperature. Higher air temperature helps for better atomization of fuel which improves fuel combustion inside the combustion chamber. But $NO_x$ emissions increase at higher CR as compared to original conditions. This is due to the higher temperature inside the combustion chamber at higher CR conditions as compared to original CR conditions. This is also due to the higher oxygen contents in the biodiesel-oil blends which improves combustion and hence increases $NO_x$ formation. The exhaust gas temperature also found to be higher due to the higher intake temperature and then better combustion rate. Similarly, Dubey and Gupta [19] had also altered CR with the aim of better performance of binary fuel blends in diesel engines. They found that the JB50TO50 gives the better result as compared to other blends and BTE efficiency improved by 2.17% a full load condition and higher CR. The tailpipe emissions CO, HC, and $NO_x$, and smoke opacity were decreased by 13.04, 17.5, 4.21, and 30.8%, respectively, while there is some increment was noticed in $CO_2$, that is, 11.04%. Overall, they stated that JB50TO50 has better option at higher CR condition as a fuel for CI engines.

Fuel IT is undoubtedly an important parameter that influences the combustion, performance, and emission characteristics of any diesel engine. Advanced IT results in increase in maximum cylinder pressure and heat release rate, while the reverse trend is noticed in the case of retarded IT. Ignition delay was found to be longer with shorter combustion duration at advanced IT as compared to that with the original and retarded ITs. BSEC at advanced IT is lower than that with the original IT. Advancing the IT results in reduced CO, HC, and particulate emission. This is due to the fact that more time is being available for the complete combustion in the case of advanced ITs. $NO_x$ emission was higher at the advanced IT in comparison to that with the original IT and lower at retarded IT compared to original and advanced ITs. This is due to the fact that at retarded IT, fuel is injected near the top dead centre (TDC) and most of the fuel burn after TDC. It causes higher amount of heat going to the exhaust which results in lowering of maximum cylinder pressure and temperature [20].

Sharma and Murugan [21] used JB50TPO50 as a fuel at altered IP conditions. Outcome revealed that the IP up to 220 bar gave good engine performance as compared to those of original IP and also 230, 240, and 250 bar. At 220 bar IP, the BTE was found to be higher. HC and CO emissions were also lower at this IP as compared to original IP at full load. The smoke opacity was lower at full load compared to original IP.

## 4. Tribological aspects associated with binary fuel blends

There are two different types of tribological studies available in the literature for binary fuel compatibility in engines, that is, short-term bench tests and long-term endurance test. Before conducting the endurance test, some bench tests on four ball tester (FBT), high frequency reciprocating rig (HFRR), and pin on disc tester (POD) were carried with binary fuel blends to endow the convenient basic information about the fuel lubricity behavior, effect of oxidative stability, and engine oil dilution. The effect of temperature and load on friction and wear were investigated with the help of FBT [4]. As biodiesel is subjected to oxidation and has highly affected lubricity at higher temperature and load, so the effect of oxidation was also studied. The effect of dilution in lubricating oil was also evaluated by means of HFRR and POD friction monitor [5]. The effect of engine oil dilution on pin and disc as well as on cam and tappet in valve train combination were studied. The effect of binary fuel blend on injection system of engine was also discussed by Singh et al. [22]. Result revealed that biodiesel and its oil blends show improved lubricity as compared to diesel, and hence increase the life of injection system of the engine.

The outcome of the four ball tester gives an idea about the impact of binary fuel blend used in the experiment on wear and friction under different operating conditions. The operating loads and temperatures were 147–392 N and 45–60–75°C, respectively. The changes in load and temperature highly influenced the wear and friction between sliding surfaces. Higher load and higher temperature adversely affect the tribological performance of the used feedstocks. The existence of both abrasive and adhesive wear has been observed at this condition, while only abrasive wear occurred at low temperature and load condition. The oxidation of methyl ester blends at higher loads and temperature conditions lead to higher corrosive wear. However, the aged (oxidized) methyl ester shows better results in terms of lubricity in short-term test [4].

The effect of 10% dilution of fresh AB70EU30 oxidized AB70EU30 and diesel by volume in the engine oil by means of HFRR and POD. The results revealed that the 10% dilution AB70EU30 showed promising results after dilution with lubricating oil as compared to diesel and oxy-biodiesel and provided much smoother mating surfaces. It is due to the existence of palmitic acid, tocopherol, etc. in the biodiesel and oil blend [23]. The lubricating oil diluted with oxy-biodiesel shows higher COF and WSD as compared to fresh biodiesel contaminated lubricating oil. This is because of complex interactions between polar molecules of anti-wear additives and biodiesel after the oxidation. This shows the poorer results as compared to fresh biodiesel but rather it is better than diesel. AB70EU30 diluted lubricating oil shows lower change in the value of total acid number (TAN) after the test as compared to diesel contaminated lubricating oil. This also justified the better condition of the lubricating oil after the dilution of AB70EU30. The ferrography also showed the less wear debris in the AB70EU30 contained lubricating oil. In modern diesel engine, additional late in-cylinder injection strategy is implemented to raise the EGT in order to check the particulate emission. During late injection, piston moves toward BDC exposing more cylinder surface area to the fuel, causing lubricating oil dilution as fuel readily passes through to the crankcase [24]. Crankcase dilution can decrease viscosity and lubricity of engine lube oil, and at the same time, it can diminish the performance of anti-wear additives. Sharma and Murugan [25] have assessed the use of 20% tyre pyrolysis oil with 80% Jatropha oil methyl ester in CI engine for long term, that is, 100 hrs. They reported that the deposition of carbon on injector tip and inside the combustion chamber was higher for this blend. But, they did not find any problem with the use of JB80TPO20 for 100 hrs in diesel engine with modified conditions.

## 5. Conclusion

Though, difference results in terms of performance and emission characteristics obtained from binary fuel blends due to their different physico-chemical properties, however, some general conclusions were drawn:

- Suitable selection and blending of two biofuels (biodiesel or oil) are an effective way to overcome the operative issues in engine associated with the neat oil and biodiesel. Typical issues such as viscosity and calorific values can be suitably adjusted with the blending.
- It can be drawn from the above observations that optimum biodiesel-oil blend suggested gives satisfactory result in terms of performance and emission.
- Minor tweaking in engine parameters can be a better option for the utilization of binary fuel blend. It provides the better efficiency and lower emissions for binary fuel blends as compared to diesel.
- The optimum blends also have better performance in short-term and long-term tribological tests as compared to diesel.
- In future, these optimum blends are environmentally friendly alternate for diesel without compromising the efficiency and the consumption of fuel as compared to diesel. Hence, these blends play a vital role to reduce the environmental impact of fossil energy sources.

## Nomenclature

| | |
|---|---|
| AB | amla biodiesel |
| ASTM | American Society for Testing Materials |
| BSFC | brake specific fuel consumption |
| bTDC | before top dead centre |
| BTE | brake thermal efficiency |
| CA | crank angle |
| CI | compression ignition |
| CO | carbon monoxide |
| CO2 | carbon dioxide |
| CR | compression ratio |
| DI | direct injection |
| EGT | exhaust gas temperature |
| EGT | exhaust gas temperature |

| | |
|---|---|
| EU | eucalyptus oil |
| GHG | greenhouse gas |
| HC | hydrocarbon |
| IC | internal combustion |
| IP | injection pressure |
| IT | injection timing |
| JB | Jatropha biodiesel |
| KME | Kapok methyl ester |
| $NO_x$ | nitrogen oxide |
| PM | particulate matter |
| TO | turpentine oil |
| TPO | tyre pyrolysis oil |

## Author details

Paramvir Singh*, S.R. Chauhan and Varun Goel

*Address all correspondence to: param@nith.ac.in

Department of Mechanical Engineering, National Institute of Technology Hamirpur, Himachal Pradesh, India

## References

[1] Singh P, Chauhan SR, Goel V. Assessment of diesel engine combustion, performance and emission characteristics fuelled with dual fuel blends. Renewable Energy. 2018;**125**:501-510. DOI: 10.1016/j.renene.2018.02.105

[2] Singh P, Chauhan SR, Varun. Carbonyl and aromatic hydrocarbon emissions from diesel engine exhaust using different feedstock: A review. Renewable and Sustainable Energy Reviews. 2016;**63**:269-291

[3] Knothe G, Razon LF. Biodiesel fuels. Progress in Energy and Combustion Science. 2017;**58**:36-59

[4] Singh P, Varun, Chauhan SR. Influence of temperature on tribological performance of dual biofuel. Fuel. 2017;**207**:751-762

[5] Singh P, Goel V, Chauhan SR. Impact of dual biofuel approach on engine oil dilution in CI engines. Fuel. 2017;**207**:680-689

[6] Peterson CL, Moller G. Biodiesel fuels: Biodegradability, biological and chemical oxygen demand, and toxicity. In: The Biodiesel Handbook. 2005. pp. 145-160

[7] Demirbas A. Progress and recent trends in biofuels. Progress in Energy and Combustion Science. 2007;**33**(1):1-8

[8] Demirbas A. Biodiesel. London: Springer; 2008

[9] Barabás I, Todoruț IA. Biodiesel quality, standards and properties. In: Biodiesel-Quality, Emissions and by-Products 2011. Rijeka: In Tech

[10] Xiao G, Gao L. First generation biodiesel. In: Biofuel Production-Recent Developments and Prospects 2011. Rijeka: In Tech

[11] Kumar N, Varun, Chauhan SR. Performance and emission characteristics of biodiesel from different origins: A review. Renewable and Sustainable Energy Reviews. 2013;**21**: 633-658

[12] Singh P, Varun, Chauhan SR, Kumar N. A review on methodology for complete elimination of diesel from CI engines using mixed feedstock. Renewable and Sustainable Energy Reviews. 2016;**57**:1110-1125

[13] Shahir SA, Masjuki HH, Kalam MA, Imran A, Fattah IR, Sanjid A. Feasibility of diesel-biodiesel-ethanol/bioethanol blend as existing CI engine fuel: An assessment of properties, material compatibility, safety and combustion. Renewable and Sustainable Energy Reviews. 2014;**32**:379-395

[14] Devan PK, Mahalakshmi NV. A study of the performance, emission and combustion characteristics of a compression ignition engine using methyl ester of paradise oil-eucalyptus oil blends. Applied Energy. 2009;**86**:675-680

[15] Vallinayagam R, Vedharaj S, Yang WM, Lee PS, Chua KJE, Chou SK. Pine oil-biodiesel blends: A double biofuel strategy to completely eliminate the use of diesel in a diesel engine. Applied Energy. 2014;**130**:466-473

[16] Sharma A, Murugan S. Investigation on the behaviour of a DI diesel engine fueled with Jatropha Methyl Ester (JME) and Tyre Pyrolysis Oil (TPO) blends. Fuel. 2013;**108**:699-708

[17] Dubey P, Gupta R. Effects of dual bio-fuel (Jatropha biodiesel and turpentine oil) on a single cylinder naturally aspirated diesel engine without EGR. Applied Thermal Engineering. 2017;**115**:1137-1147

[18] Sharma A, Murugan S. Potential for using a tyre pyrolysis oil-biodiesel blend in a diesel engine at different compression ratios. Energy Conversion and Management. 2015;**93**:289-297

[19] Dubey P, Gupta R. Influences of dual bio-fuel (Jatropha biodiesel and turpentine oil) on single cylinder variable compression ratio diesel engine. Renewable Energy. 2018;**115**: 1294-1302

[20] Sharma A, Murugan S. Combustion, performance and emission characteristics of a DI diesel engine fuelled with non-petroleum fuel: A study on the role of fuel injection timing. Journal of the Energy Institute. 2015;**88**(4):364-375

[21] Sharma A, Sivalingam M. Impact of fuel injection pressure on performance and emission characteristics of a diesel engine fueled with Jatropha methyl ester tyre pyrolysis blend. In: SAE Technical Paper; 2014

[22] Singh P, Goel V, Chauhan SR. Effects of dual biofuel approach for total elimination of diesel on injection system by reciprocatory friction monitor. Proceedings of the Institution of Mechanical Engineers, Part J: Journal of Engineering Tribology. 2018;**232**(9):1068-1076. DOI: 10.1177/1350650117737874

[23] Kumar N, Varun, Chauhan S. Analysis of tribological performance of biodiesel. Proceedings of the Institution of Mechanical Engineers, Part J: Journal of Engineering Tribology. 2014;**228**(7):797-807

[24] Kumar N, Varun, Chauhan SR. Evaluation of endurance characteristics for a modified diesel engine runs on jatropha biodiesel. Applied Energy. 2015;**155**:253-269

[25] Sharma A, Murugan S. Durability analysis of a single cylinder DI diesel engine operating with a non-petroleum fuel. Fuel. 2017;**191**:393-402

Chapter 9

# Comparison of Ethanol and Methanol Blending with Gasoline Using Engine Simulation

Simeon Iliev

Additional information is available at the end of the chapter

http://dx.doi.org/10.5772/intechopen.81776

**Abstract**

During the last years, concerns regarding climate change, decline of energy security, and hydrocarbon reserves have resulted in a wide interest in renewable alternative sources for transportation fuels. Methanol and ethanol have been possible candidates as alternative fuels for the internal combustion engines because they are liquid and have several physical and combustion properties which resemble those of gasoline. Therefore, the aim of this study is to develop the one-dimensional model of a gasoline engine for predicting the effect of various fuel types on engine performances, specific fuel consumption, and emissions. Commercial software AVL BOOST was used to examine the engine characteristics for different blends of methanol, ethanol, and gasoline (by volume). A comparison was made between the results gained from the engine simulation of different fuel blends and those of gasoline. They show that when blended fuel was used, the engine brake power decreased and the BSFC increased compared to those of gasoline fuel. When blended fuel increases, the CO and HC emissions decrease, and there is a major increase in NOx emissions when blended fuel increases up to 30% M30 (E30). Increase in the percentage of ethanol and methanol leads to a significant increase in NOx emissions.

**Keywords:** alternative fuels, ethanol blends, methanol blends, engine simulation, spark-ignition engine, emissions

## 1. Introduction

In the last years, the problem with crude oil depletion has arisen. Intensive research has been carried out to find out alternative to fossil fuels. Alternative fuels are derived from resources different from petroleum. When used in internal combustion engines (ICE), these fuels generate lower air pollutants compared to petrol fuel, and a majority of them are more economically beneficial compared to fossils fuels. They are also renewable. The most common fuels that are used as alternative fuels are natural gas, propane, methanol, ethanol, and hydrogen. Regarding engine operating with blended fuels, a lot of papers have been written about these blended fuels; but a small number of works have compared some of these fuels together in the same engine [1–4]. Low contents of ethanol or methanol have been added to gasoline since at least the 1970s, when there was a reduction in oil supplies and scientists began searching for alternative energy carriers in order to replace petrol fuels. In the beginning, ethanol and methanol were thought to be the most attractive alcohols to be added to gasoline. Ethanol and methanol can be manufactured from natural products or waste materials, whereas gasoline fuel which is a nonrenewable energy resource cannot be manufactured [5, 6]. An important feature is that methanol and ethanol can be used without requiring any significant changes in the structure of the engine. Being part of the various alcohols, ethanol and methanol are known as the most suitable fuels for spark-ignition (SI) engines.

The use of blended fuels is crucial since many of these blends can be used in engines with the aim to improve its performance, efficiency, and emissions. The oxygenates are one of the most important fuel additives to improve fuel efficiency (organic oxygen-containing compounds). A few oxygenates have been used as fuel additives, such as ethanol, methanol, methyl tertiary butyl alcohol, and tertiary butyl ether [7]. The process of using oxygenates makes more oxygen available in the combustion process and has a great potential to reduce SI engine exhaust emissions.

Regarding the combustion process, the flash point and autoignition temperature of methanol and ethanol are higher than pure gasoline, which makes it safer for storage and transportation. The latent heat of ethanol of evaporation is three to five times higher than pure gasoline; this leads to increase the volumetric efficiency because temperature of the intake manifold is lower. The heating value of ethanol is lower than gasoline. Consequently, 1.6 times more alcohol fuel is needed to achieve the exact same energy output. The stoichiometric air-fuel ratio of ethanol is around two-third of the pure gasoline; therefore, for complete combustion, the needed amount of air is lesser for ethanol [8]. Ethanol has several advantages compared to gasoline, e.g., lowering of unburned HC emissions, CO, and much better antiknock characteristics [9]. Ethanol and methanol have a lot higher octane number compared to pure gasoline fuel [10]. This enables higher compression ratios of engines and, as a result, increases its thermal efficiency [11]. The production of methanol can be from natural gas at no great cost and is easy to blend with gasoline fuel. These properties of methanol make it as an attractive additive. Methanol is aggressive to some materials, like plastic components and some of the metals in the fuel system. When using methanol it is necessary that precautions had to be taken when handling it [12].

There are many publications with different blends of alcohols and gasoline fuel. For example, Palmer [13] examined the influence of blends of ethanol and gasoline in spark-ignition engine. The obtained results pointed out that ethanol addition (10%) leads to 5% increase in the engine power and 5% octane number increase for each 10% ethanol added. The result showed that 10% of ethanol addition to gasoline fuel lead to reduction the emissions of CO up to 30%. In another study, Bata et al. [9] examined different blends of ethanol and gasoline and discovered that ethanol reduced the UHC and CO emissions. The lowered CO emissions are caused by the oxygenated characteristic and wide flammability of ethanol. Other researchers [14] studied that the potentialities for ethanol production are equivalent to about 32% of the total gasoline consumption worldwide, when used in 85% ethanol in gasoline for a passenger vehicle. In another study, Shenghua et al. [15] examined a gasoline engine with various percentages of methanol blends (from 10 to 30%) in gasoline. The results showed that engine torque and power decreased, whereas the brake thermal efficiency improved with the increase of methanol percentage in the fuel blend. Other authors [16] have studied the influence of methanol-gasoline blends on the gasoline engine performance. The results showed that the highest brake mean effective pressure (BMEP) was obtained from 5% methanol-gasoline blend. In another study, Altun et al. [17] researched the influence of 5 and 10% methanol and ethanol blending in gasoline fuel on engine performance and emissions. The best result in emissions showed blended fuels. The HC emissions of E10 and M10 are reduced by 13 and 15% and the CO emissions by 10.6 and 9.8%, respectively. An increased $CO_2$ emission for E10 and M10 was observed. The methanol and ethanol addition to gasoline showed an increase in the brake-specific fuel consumption (BSFC) and a decrease in break thermal efficiency compared to gasoline.

It can be seen in the literature survey that the exhaust emissions for ethanol-gasoline and methanol-gasoline blends are lower than that of pure gasoline fuel [9, 13, 14, 17]. The engine performance and exhaust emissions with ethanol-gasoline blends resemble those with methanol-gasoline blends.

From the reviewed literature, a conclusion was made that the exhaust emission and engine performance of various blends of methanol and ethanol in gasoline engines have not been investigated sufficiently. Therefore, the objective of this work is to investigate the effects of methanol-gasoline and ethanol-gasoline fuel blends on the performance and exhaust emissions of a gasoline engine under various engine speeds, comparing them with those of pure gasoline.

The simulation tools are the most used in recent years owing to the continuous increase in computational power. The use of engine simulations enables optimization of engine combustion, geometry, and operating characteristics toward improving specific fuel consumption and exhaust emissions and reducing engine development time and costs. Consequently, it can be expected that the use of engine simulations during engine construction will continue to increase. Engine modeling is a fruitful research area, and therefore many laboratories have their own engine thermodynamic models with varying degrees of complexity, scope, and ease to use [18].

Computer simulation is becoming an important tool for time and cost efficiency in engine's development. The simulation results are challenging to be obtained experimentally. Using computational fluid dynamics (CFD) has allowed researchers to understand the flow behavior and quantify important flow parameters such as mass flow rates or pressure drops, under the condition that the CFD tools have been properly validated against experimental results. Many processes in the engine are three-dimensional; however, it requires greater knowledge and large computational time. Thus, simplified one-dimensional simulation is occasionally used. Hence, simulating the complex components by means of a three-dimensional code and modeling the rest of the system with a one-dimensional code are the right choice to save computational time, i.e., the ducts. This way, a coupling methodology between the one-dimensional and the three-dimensional codes in the respective interfaces is necessary and has become the aim of numerous authors [19–21].

## 2. Research methodology

The aim of the present chapter is to develop the one-dimensional model of four-stroke port fuel injection (PFI) gasoline engine for predicting the effect of methanol-gasoline (M0–M50) and ethanol-gasoline (E0–E50) addition to gasoline on the exhaust emissions and performance of gasoline engine. For this, simulation of gasoline SI engine (calibrated) as the basic operating condition and the laminar burning velocity correlations of methanol-gasoline and ethanol-gasoline blends for calculating the changed combustion duration was used. The engine power, specific fuel consumption, and exhaust emissions were compared and discussed [22, 23].

### 2.1. Simulation setup

The one-dimensional SI engine model is created by using the AVL BOOST software and has been employed to examine the performance and emissions working on gasoline, ethanol-gasoline, and methanol-gasoline blends.

In **Figure 1**, PFIE symbolizes the engine, while C1–C4 is the number of cylinders of the SI engine. The cylinders are the main element in this model, because they have many very important parameters to settle: the internal geometry, bore, stroke, connecting rod, length and compression ratio, as well as the piston pin offset and the mean crankcase pressure. The measuring points are marked with MP1–MP18. PL1–PL4 symbolizes the plenum. System boundary stands for SB1 and SB2. CL1 represents the cleaner. R1–R10 stands for flow restrictions. CAT1 symbolizes catalyst and fuel injectors—I1–I4. The flow pipes are numbered 1–34.

The calibrated gasoline engine model was described by Iliev [23], and its layout is shown in **Figure 1** with engine specification shown in **Table 1**.

**Table 2** presents a comparison between the properties of gasoline, ethanol, and methanol. As shown in **Table 2**, compared with gasoline and ethanol, methanol has a higher elemental oxygen content and a lower heating value, molecular weight, elemental carbon, hydrogen content, and stoichiometric air/fuel ratio (AFR).

### 2.2. Combustion model description

In this research, two-zone model of Vibe was chosen for the combustion simulation and analysis. The combustion chamber was divided into two regions: unburned gas region and burned gas regions [17]. For the burned charge and unburned charge, the first law of thermodynamics is applied:

$$\frac{dm_b u_b}{d\alpha} = -p_c \frac{dV_b}{d\alpha} + \frac{dQ_F}{d\alpha} - \sum \frac{dQ_{Wb}}{d\alpha} + h_u \frac{dm_b}{d\alpha} - h_{BB,b} \frac{dm_{BB,b}}{d\alpha} \tag{1}$$

$$\frac{dm_u u_u}{d\alpha} = -p_c \frac{dV_u}{d\alpha} - \sum \frac{dQ_{Wu}}{d\alpha} - h_u \frac{dm_B}{d\alpha} - h_{BB,u} \frac{dm_{BB,u}}{d\alpha} \tag{2}$$

http://dx.doi.org/10.5772/intechopen.81776

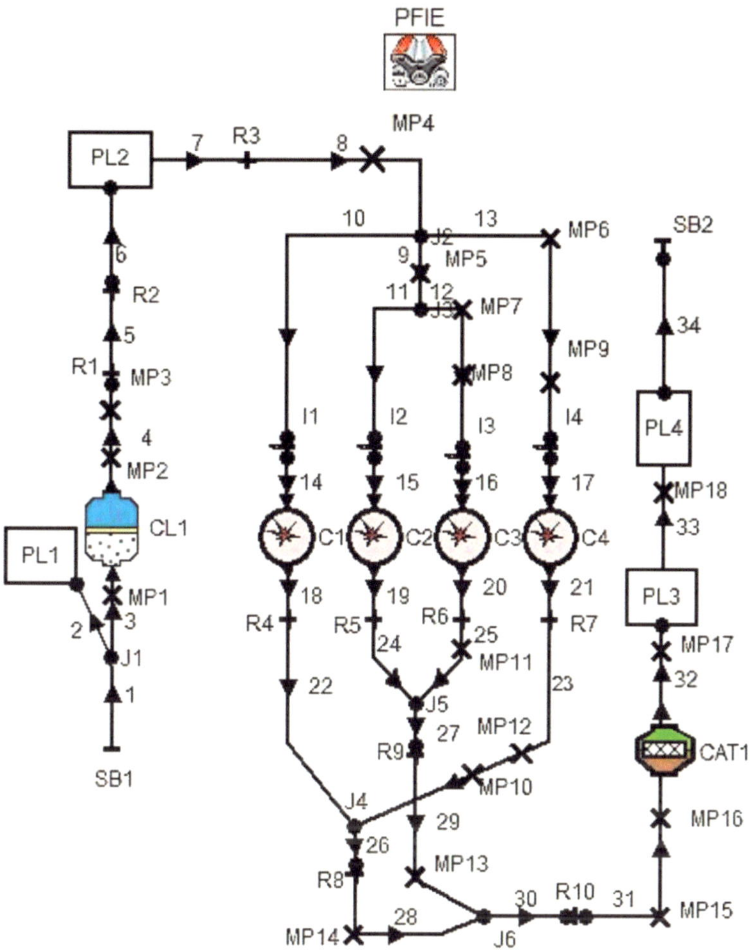


**Figure 1.** Schematic of the gasoline PFI engine model.

where $dm_u$ represents the change of the internal energy in the cylinder, $p_c \frac{dV}{da}$ is the piston work, $\frac{dQ_F}{da}$ stands for the fuel heat input, $\frac{dQ_W}{da}$ is the wall heat loses, $h_u \frac{dm_b}{da}$ represents the enthalpy flow from the unburned to the burned zone due to the conversion of a fresh charge to combustion products. The heat flux between the two zones is neglected. $h_{BB} \frac{dm_{BB}}{da}$ is the enthalpy flow due to blow-by, **u** and **b** in the subscript are unburned and burned gas.

Moreover, the sum of the zone volumes must be equal to the cylinder volume, and the sum of the volume changes must be equal to the cylinder volume change:

$$\frac{dV_b}{d\alpha} + \frac{dV_u}{d\alpha} = \frac{dV}{d\alpha} \tag{3}$$

$$V_b + V_u = V \tag{4}$$

| Engine parameters | Value |
|---|---|
| Bore | 86 (mm) |
| Stroke | 86 (mm) |
| Compression ratio | 10.5 |
| Connection rod length | 143.5 (mm) |
| Number of cylinder | 4 |
| Piston pin offset | 0 (mm) |
| Displacement | 2000 (cc) |
| Intake valve open | 20 BTDC (deg) |
| Intake valve close | 70 ABDC (deg) |
| Exhaust valve open | 50 BBDC (deg) |
| Exhaust valve close | 30 ATDC (deg) |
| Piston surface area | 5809 ($mm^2$) |
| Cylinder surface area | 7550 ($mm^2$) |
| Number of stroke | 4 |

**Table 1.** Engine specification.

| Properties | Gasoline | Methanol | Ethanol |
|---|---|---|---|
| Chemical formula | $C_8H_{15}$ | $CH_3OH$ | $C_2H_5OH$ |
| Molecular weight | 111.21 | 32.04 | 46.07 |
| Oxygen content (wt%) | — | 49.93 | 34.73 |
| Carbon content (wt%) | 86.3 | 37.5 | 52.2 |
| Hydrogen content (wt%) | 24.8 | 12.5 | 13.1 |
| Stoichiometric AFR | 14.5 | 6.43 | 8.94 |
| Lower heating value (MJ/kg) | 44.3 | 20 | 27 |
| Heat of evaporation (kJ/kg) | 305 | 1.178 | 840 |
| Research octane number | 96.5 | 112 | 111 |
| Motor octane number | 87.2 | 91 | 92 |
| Vapor pressure (psi at 37.7 OC) | 4.5 | 4.6 | 2 |
| Destiny (g/$cm^3$) | 0.737 | 0.792 | 0.785 |
| Normal boiling point (OC) | 38–204 | 64 | 78 |
| Autoignition temperature (OC) | 246–280 | 470 | 365 |

**Table 2.** Comparison of fuel properties.

The amount of burned mixture at each time setup is obtained from the Vibe function. For all other terms, for instance, wall heat losses, etc., models similar to the single-zone models with an appropriate distribution on the two zones are used [24].

### 2.3. A description of exhaust emission model

In AVL BOOST, the model of formation on NOx is based on AVL List Gmbh [24], which incorporates the Zeldovich mechanism [25]. The rate of NOx production was obtained using Eq. (5):

$$r_{NO} = C_{PPM}\, C_{KM}(2, 0).(1 - \alpha^2).\left(\frac{r_1}{1 + \alpha AK_2} + \frac{r_4}{1 + AK_4}\right). \tag{5}$$

where $\alpha = \frac{C_{NO.act}}{C_{NO.equ}} \cdot \frac{1}{C_{KM}},\ AK_2 = \frac{r_1}{r_2 + r_3},\ AK_4 = \frac{r_4}{r_5 + r_6}.$

In the above equation, $C_{PPM}$ represents post-processing multiplier, $C_{KM}$ denotes kinetic multiplier, $C$ stands for molar concentration in equilibrium, and $r_i$ represents reaction rates of Zeldovich mechanism.

The NOx formation model in AVL Boost is based on Onorati et al. [26]:

$$r_{CO} = C_{Const}(r_1 + r_2).(1 - \alpha) \tag{6}$$

$$\text{where } \alpha = \frac{C_{CO.act}}{C_{CO.equ}}.$$

In Eq. (6), $C$ represents molar concentration in equilibrium and $r_i$ represents reaction rates based on the model.

The unburned HC has different sources. A complete description of HC formation still cannot be given, and the achievement of a reliable model within a thermodynamic approach is definitely prevented by the fundamental assumptions and the requirement of reduced computational times. Still, a phenomenological model which accounts for the main formation mechanisms and is able to capture the HC trends as function of the engine operating parameter may be proposed. The following important sources of unburned HC can be identified in SI engines [21]:

1. During the intake and compression stroke, fuel vapor is absorbed into the oil layer and deposits on the cylinder walls. The following desorption occurs when the cylinder pressure decreases during the expansion stroke and complete combustion cannot take place anymore.
2. A fraction of the charge enters the crevice volumes and is not burned since the flame quenches at the entrance.
3. Occasional complete misfire or partial burning takes place when combustion quality is poor.
4. Quench layers on the combustion chamber wall which are left as the flame extinguishes prior to reaching the walls.

5. The flow of fuel vapor into the exhaust system during valve overlap in gasoline engines.

The first two mechanisms and in particular the crevice formation are considered to be the most important and need to be accounted for in a thermodynamic model. Partial burn and quench layer effect cannot be physically described in a quasi-dimensional approach, but may be included by adopting tunable semiempirical correlations.

The formation of unburned HC in the crevices is described by assuming that the pressure in the cylinder and in the crevices is the same and that the temperature of the mass in the crevice volumes is equal to the piston temperature.

The mass in the crevices at any time is described by Eq. (7):

$$m_{crevice} = \frac{pV_{crevice}M}{RT_{piston}} \tag{7}$$

In Eq. (7), $m_{crevice}$ represents the mass of unburned charge in the crevice, $p$ denotes cylinder pressure, $V_{crevice}$ stands for total crevice volume, $M$ represents unburned molecular weight, $T_{piston}$ is the temperature of the piston, and $R$ denotes gas constant.

The second important source of HC is the presence of lubricating oil in the fuel or on the walls of the combustion chamber. During the compression stroke, the fuel vapor pressure increases so, by Henry's law, absorption occurs even if the oil was saturated during the intake. During combustion the concentration of fuel vapor in the burned gases goes to zero so the absorbed fuel vapor will desorb from the liquid oil into the burned gases. Fuel solubility is a positive function of the molecular weight, so the oil layer contributed to HC emissions depending on the different solubility of individual hydrocarbons in the lubricating oil.

The assumptions made in the development of the HC absorption/desorption are the following:

1. Fuel is constituted by a single hydrocarbon species, completely vaporized in the fresh mixture.
2. The oil film temperature is at the same as the cylinder wall.
3. Traverse flow across the oil film is negligible.
4. Oil is represented by squalane ($C_{30}H_{62}$), whose characteristics resemble those of the SAE5W20 lubricant.
5. Diffusion of the fuel in the oil film is the limiting factor, for the diffusion constant in the liquid phase which is 104 times smaller than the corresponding value in the gas phase.

The radial distribution of the fuel mass fraction in the oil film can be determined by solving the diffusion Eq. (8):

$$\frac{\partial w_F}{\partial t} - D\frac{\partial^2 w_F}{\partial r^2} = 0 \tag{8}$$

In Eq. (8), $w_F$ represents fuel's mass fraction in the oil film, $t$ is the time, $r$ stands for radial position in the oil film (distance from the wall), and $D$ is relative (fuel-oil) diffusion coefficient.

## 3. Result and discussion

The present research focused on the performance and emission characteristics of the methanol and ethanol-gasoline blends. Various concentrations of the blends 0% methanol (ethanol) M0 (E0), 5% methanol (ethanol) M5 (E5), 10% methanol (ethanol) M10 (E10), 20% methanol (ethanol) M20 (E20), 30% methanol (ethanol) M30 (E30), 50% methanol (ethanol) M50 (E50), and 85% methanol (ethanol) M85 (E85) by volume were analyzed.

### 3.1. Engine performance characteristics

The results of the brake power and specific fuel consumption for ethanol-gasoline blended fuels at different engine speeds are shown on **Figures 2** and **3**.

The brake power is one of the important factors that determine the performance of an engine. The variation of brake power with speed was obtained at full load conditions for E5, E10, E20, E30, E50, and pure gasoline E0. The ethanol content in the blended fuel increased, and the brake power decreased for all engine speeds. The gasoline brake power was higher than E5–E50 for all engine speeds. The ethanol's heat of evaporation is higher in comparison to gasoline fuel, providing air-fuel charge cooling and increasing the density of the charge. The blended fuel causes the equivalence ratio of blend approaches to stoichiometric condition which can lead to a better combustion. However, the ethanol heating value is lower compared to gasoline, and it can neutralize the previous positive effects. Consequently, a lower power output is obtained.

**Figure 3** shows the changes of the BSFC for ethanol-gasoline blends under various engine speeds. The figure shows that the BSFC increased as the ethanol percentage increased. Heating value and stoichiometric air-fuel ratio are the smallest for these two fuels, which means that for specific air-fuel equivalence ratio, more fuel is needed. The highest specific fuel consumption is obtained at E50 ethanol-gasoline blend.

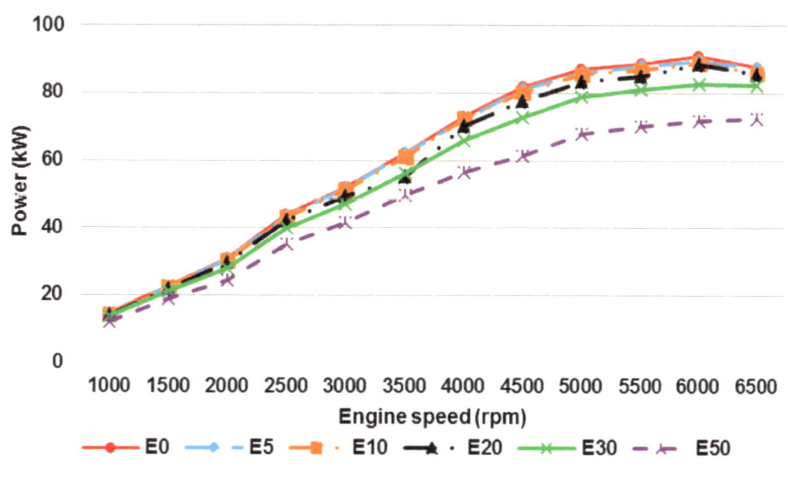


**Figure 2.** Influence of ethanol-gasoline blended fuels on brake power.

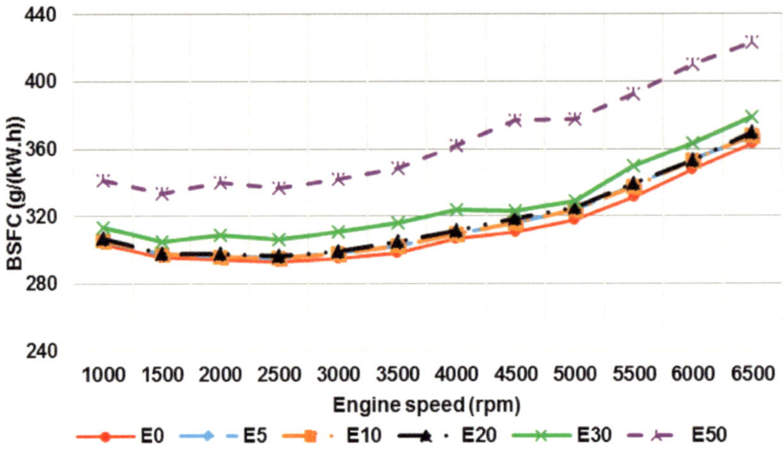


**Figure 3.** Influence of ethanol-gasoline blended fuels on brake-specific fuel consumption.

Moreover, there is a slight difference between the BSFC when using pure gasoline and when using blends (E5, E10, and E20). The lower energy content of blended fuels causes some increment in BSFC of the engine.

**Figure 4** shows the influence of methanol-gasoline blended fuels on engine brake power. The variation of brake power with speed was obtained at full load conditions for M5, M10, M20, M30, M50, and pure gasoline M0. When the methanol content in the blended fuel was increased (M10, M20, and M30), there was not a significant increase in engine brake power.

The engine brake power may be due to the increase of the indicated mean effective pressure for higher methanol content blends. The methanol's heat of evaporation is higher compared to that of gasoline, thus providing air-fuel charge cooling and increasing the density of the charge. Therefore, a higher power output is obtained. The engine brake power was higher in operation with gasoline in comparison to M50 for all engine speeds.

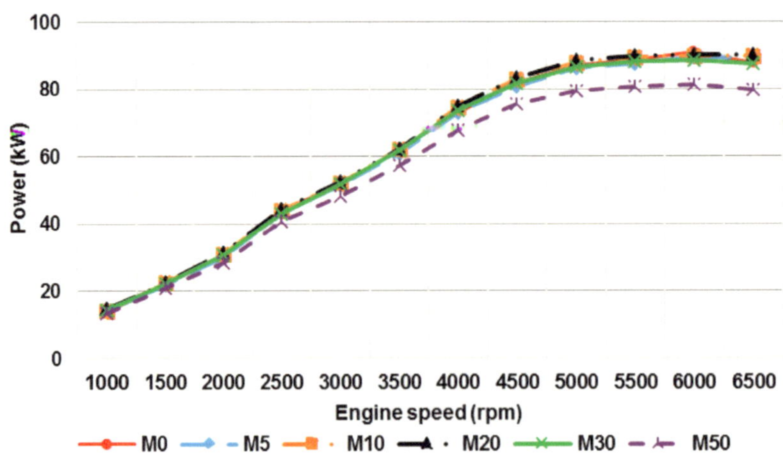


**Figure 4.** Influence of methanol-gasoline blended fuels on brake power.

**Figure 5** shows the variations of the BSFC for methanol-gasoline blended fuels under various engine speeds. As shown in this figure, the BSFC increased as the methanol percentage increased. This can be described with heating value, and stoichiometric air-fuel ratio is the smallest for these two fuels, which means that for specific air-fuel equivalence ratio, more fuel is needed. The specific fuel consumption of M50 methanol-gasoline blend was highest compared to those of gasoline for all engine speeds.

Furthermore, there is a small difference between the BSFC when using gasoline and when using methanol-gasoline blended fuels (M5–M30). As engine speed increased reaching 2000 rpm, the BSFC decreased reaching its minimum value.

The results of the brake power and specific fuel consumption for ethanol- and methanol-gasoline blended fuels at different engine speeds are presented in **Figures 6** and **7**.

When there was an increase in the ethanol content in the blended fuel, the brake power decreased for all engine speeds. The brake power of gasoline fuel was higher than those of E5–E50. The heating value of ethanol is lower than pure gasoline fuel, and the heating value of the blends decreases with the increase of the ethanol percentage. Consequently, a lower power output is obtained [22, 23].

By increasing the percentage of methanol in the blends (M5 and M10), the brake power slightly increased, which can be explained by better combustion efficiency of oxygenated fuels. By increasing the methanol content in the blends (M30 and M50), the engine brake power decreased for all engine speeds. The blended fuel heating value decreases with the increase of the percentage of methanol. This results in a lower power output. The gasoline brake power was higher compared to blend M50.

**Figure 7** shows the changes of the BSFC for blended fuels under different engine speeds. The BSFC increased as the ethanol and methanol percentage increased. The reason has been known—the heating value and stoichiometric air-fuel ratio are the smallest for this fuel, which

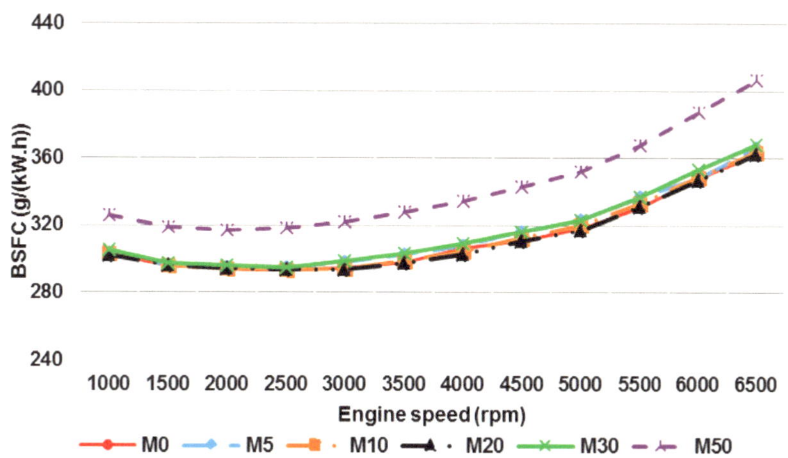


**Figure 5.** Influence of methanol-gasoline blended fuels on engine brake power.

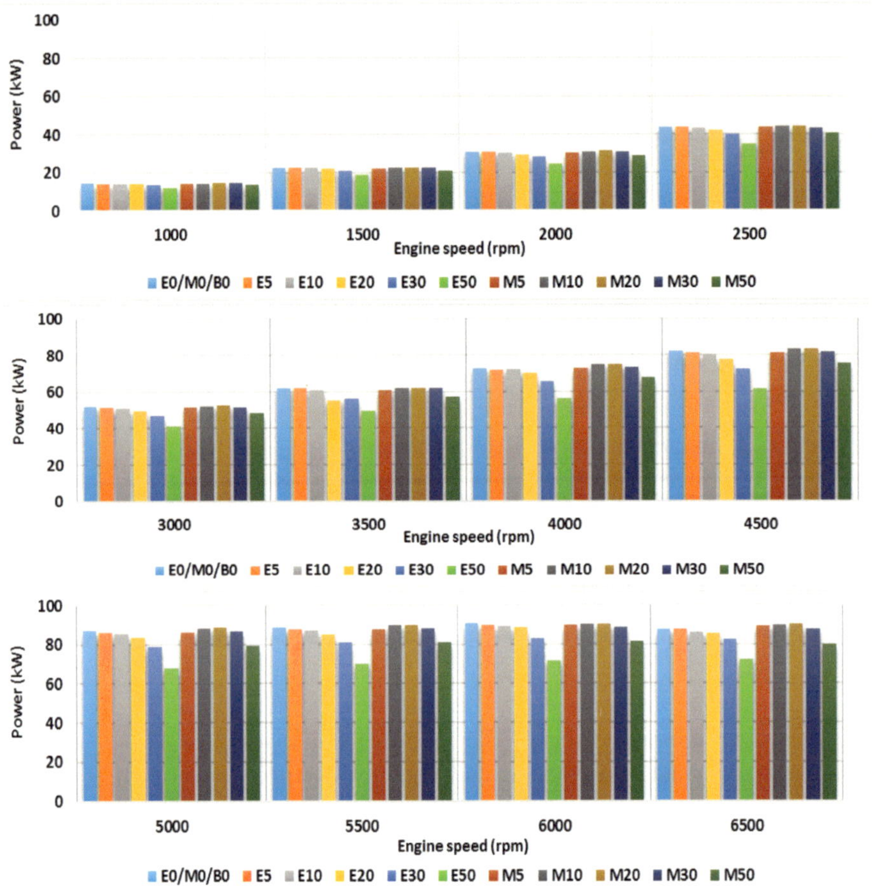


**Figure 6.** Effect of blended fuels on engine brake power.

means that more fuel is needed for specific air-fuel equivalence ratio. The highest specific fuel consumption is obtained at E50 (M50) blended fuel.

What is more, there is small difference between the BSFC when using pure gasoline and blended fuels (E5 (M5), E10 (M10), and E20 (M20)). The lower energy content of ethanol blended fuels makes some increment in BSFC.

### 3.2. Emission characteristics

The result of the ethanol-blended fuels on CO emissions is shown in **Figure 8**.

A conclusion, which can be made by **Figure 8**, is that when ethanol content increases, the CO emission decreases. The reason for this could be explained with the enrichment of oxygen owing to the ethanol, in which an increase in the proportion of oxygen will promote the further oxidation of CO during the engine exhaust process. One of the other significant reasons for this reduction is that ethanol ($C_2H_5OH$) has less carbon than gasoline ($C_8H_{18}$).

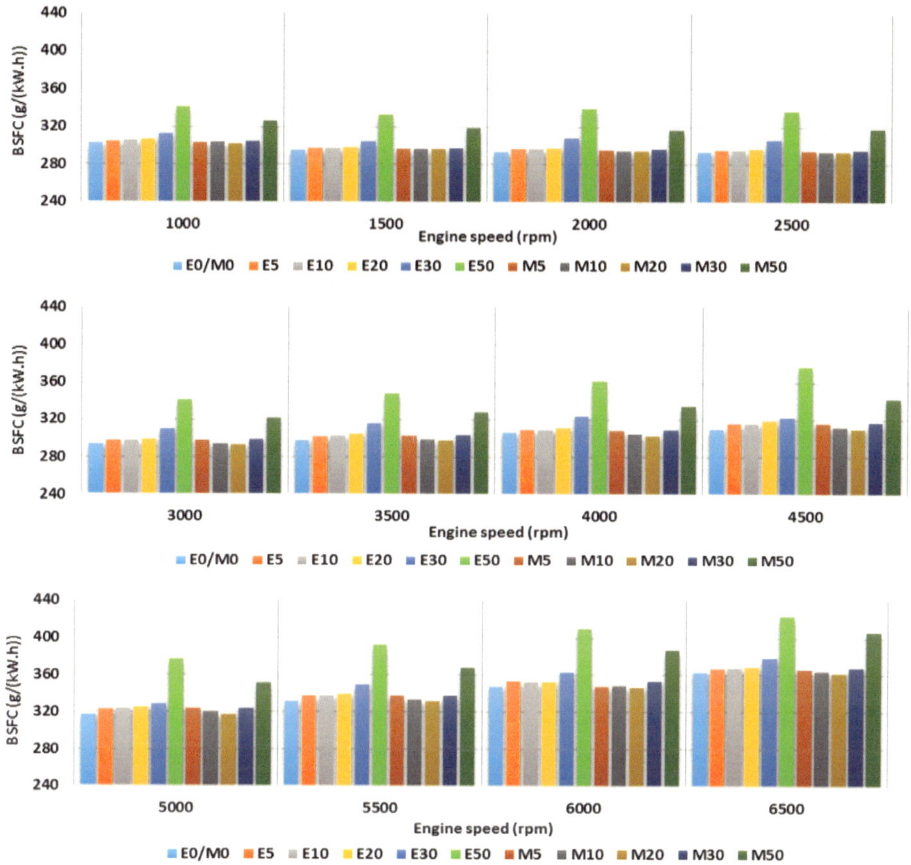


**Figure 7.** Influence of blended fuels on engine fuel consumption.

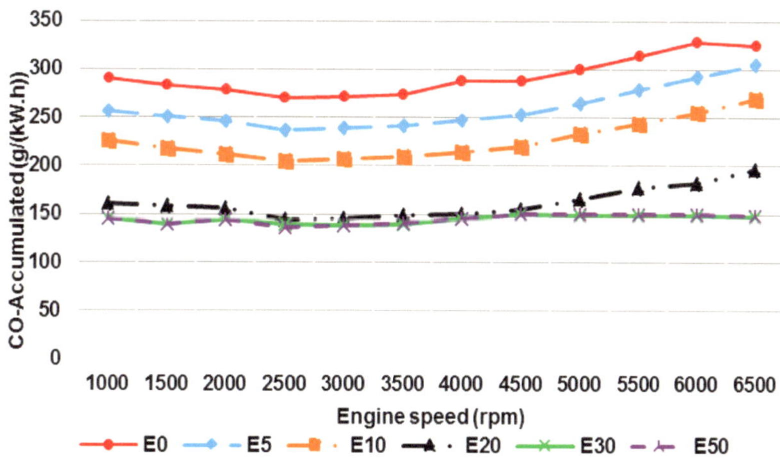


**Figure 8.** Influence of ethanol-gasoline blended fuels on CO emissions.

The result of the ethanol gasoline blends on HC emissions is shown in **Figure 9.** The figure shows that when ethanol percentage increases, the HC concentration decreases. The HC emission decreases with the increase of the relative air-fuel ratio. The decrease of HC can be explained similarly to that of CO concentration described above.

The effect of the ethanol gasoline blends on NOx emissions for various engine speeds is shown in **Figure 10**. When the combustion process is closer to stoichiometric, flame temperature increases. As a result, the NOx emissions are increased.

The effect of the methanol-gasoline blends on CO emissions for various engine speeds can be seen in **Figure 11**. When methanol percentage increases, the CO concentration decreases. This can be explained with the enrichment of oxygen because of the methanol and less carbon of methanol than gasoline.

The effect of the methanol-gasoline blends on HC emissions is visible in **Figure 12**. When methanol percentage increases, the HC concentration decreases. The concentration of HC emissions decreases with the increase of the relative air-fuel ratio. The reason for the decrease of HC concentration resembles that of ethanol.

The effect of the methanol-gasoline blends on NOx emissions can be seen in **Figure 13**. When methanol percentage increases, the NOx concentration increases. When combustion process is closer to stoichiometric, flame temperature increases and the NOx emissions increase as well.

The effect of the ethanol- and methanol-gasoline blends on CO emissions can be viewed in **Figure 14**. By increasing the methanol and ethanol content in the blended fuel, the CO emission decreases. The reason can be the enrichment of oxygen because of the ethanol and methanol, in which an increase in the proportion of oxygen will promote the further oxidation of CO during the engine exhaust process. Another major reason for this reduction is that ethanol ($C_2H_5OH$) and methanol ($CH_3OH$) have less carbon than gasoline ($C_8H_{18}$). The lowest CO emissions are obtained with blended fuel containing methanol (M50).

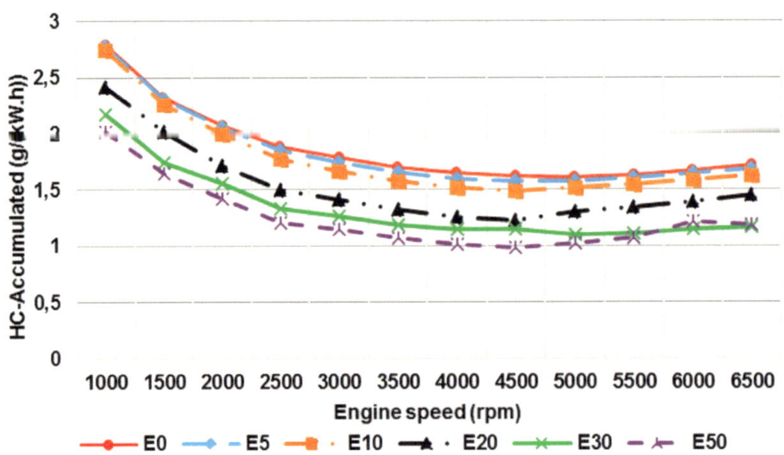


**Figure 9.** Influence of ethanol-gasoline blended fuels on HC emissions.

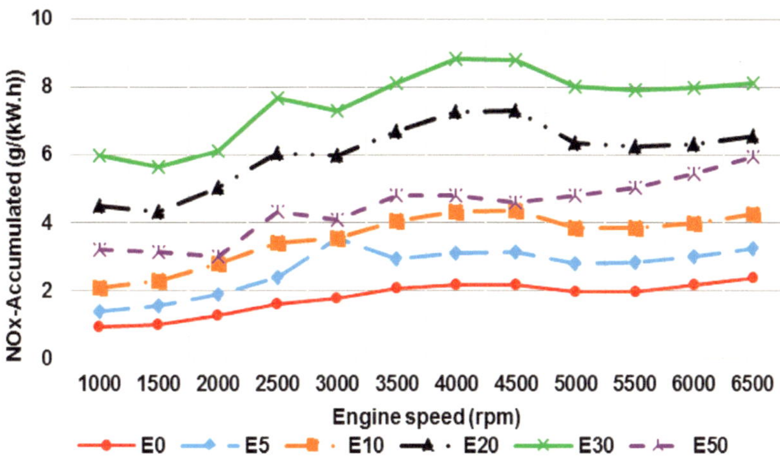


**Figure 10.** Influence of ethanol-gasoline blended fuels on NOx emissions.

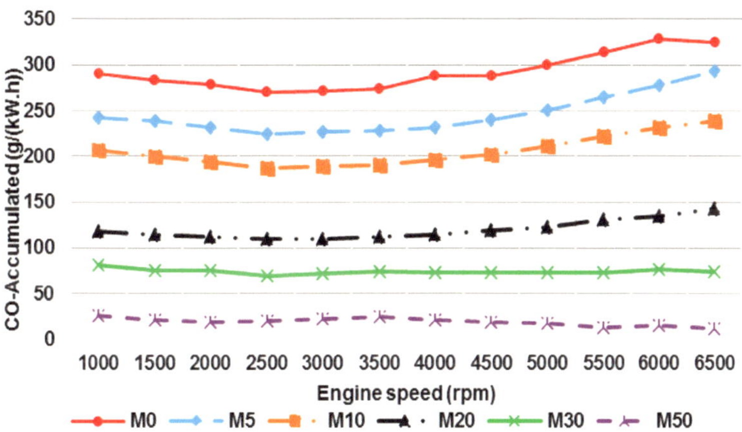


**Figure 11.** Influence of methanol-gasoline blended fuels on CO emissions.

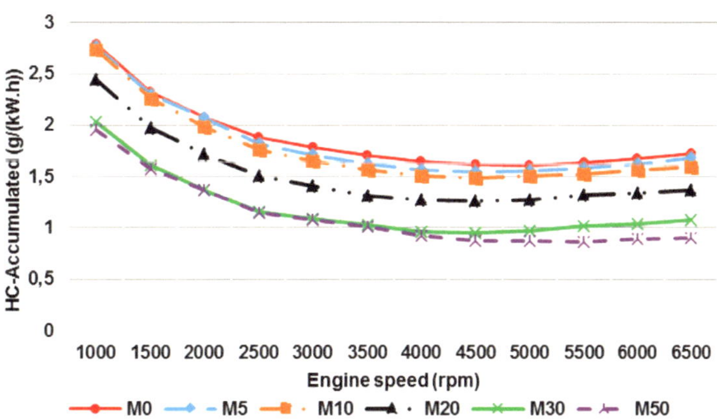


**Figure 12.** Influence of methanol-gasoline blended fuels on HC emissions.

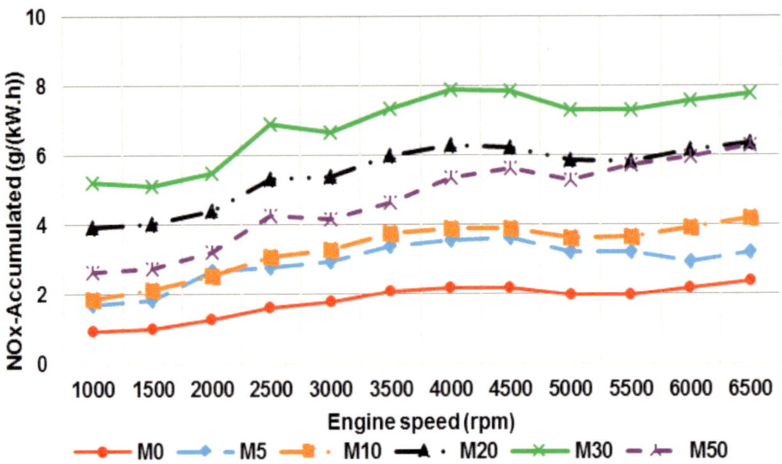


Figure 13. Influence of methanol-gasoline blended fuels on NOx emissions.

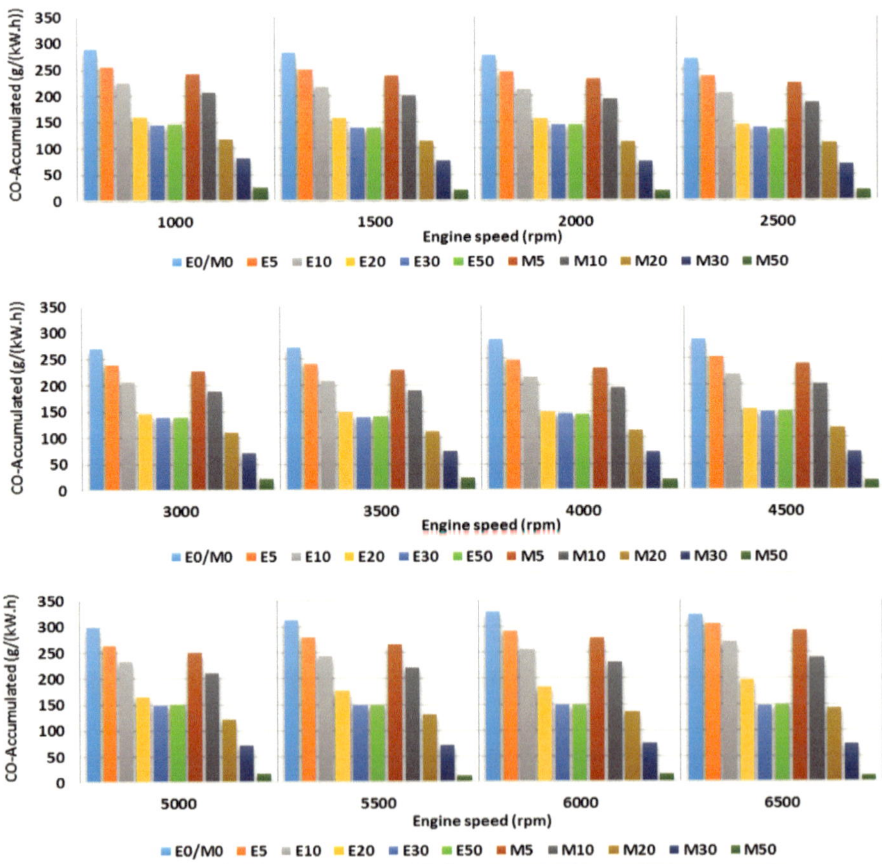


Figure 14. Influence of ethanol- and methanol–gasoline blended fuels on CO emissions.

The effect of the ethanol- and methanol-gasoline blends on HC emissions is visible in **Figure 15**. When there is an increase in the ethanol and methanol percentage, the HC concentration decreases.

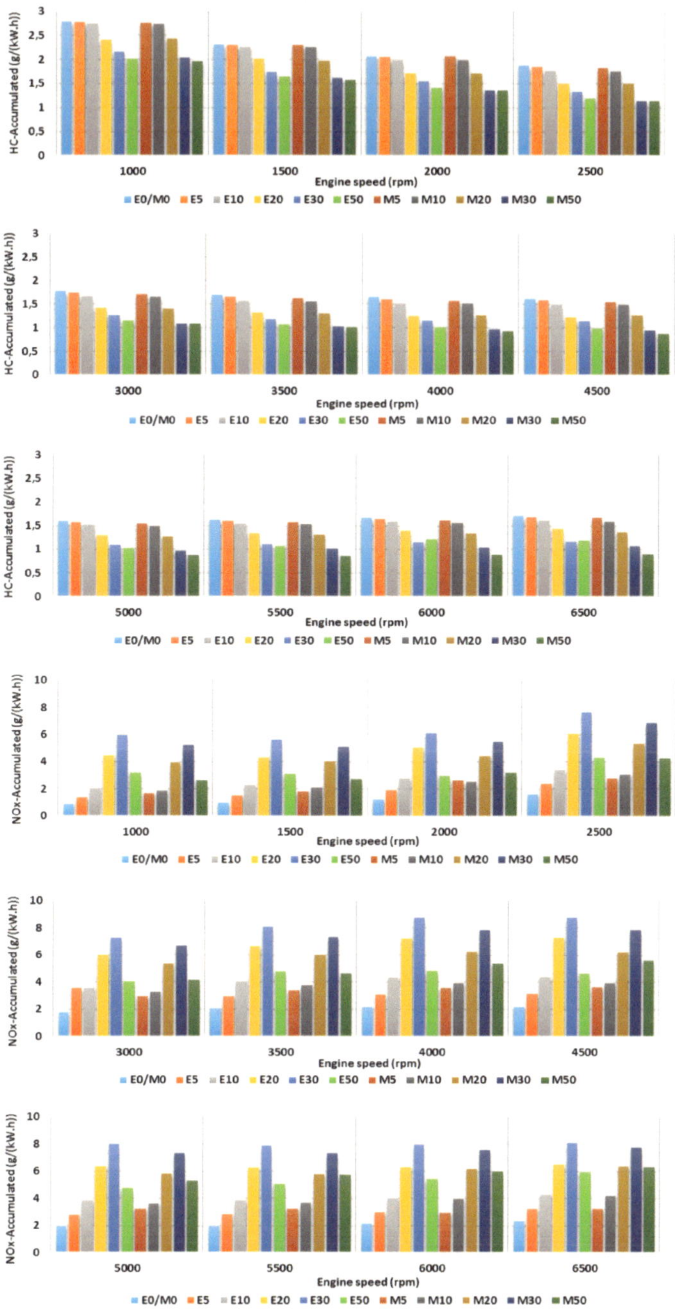


**Figure 15.** Influence of blended fuels on HC and NOx emissions.

When the relative air-fuel ratio increases, the concentration of HC emissions decreases. The reason for the decrease in HC emissions is similar to that of CO described above. The comparison between the decrease in HC emissions and the blended fuels indicates that methanol is more effective than ethanol. The lowest HC emissions are obtained with methanol-blended fuel (M50). When more combustion is complete, it will result in lower HC emissions.

**Figure 15** shows the influence of the blended fuels on NOx emissions. It is noticeable that when methanol and ethanol percentage increases up to 30% E30 (M30), the NOx emission increases, after which it decreases with increasing the percentage of the methanol (ethanol).

The reason is that the improved combustion results in increased temperature in combustion chamber. The higher methanol (ethanol) content in the blends lowers the temperature in combustion chamber. The lower temperature is due to:

1. Latent heat of evaporation of alcohols, which decreases the temperature in combustion chamber during the vaporization.
2. The more triatomic molecules are produced: the higher the gas heat capacity and the lower the combustion gas temperature will be. However, the low temperature in combustion chamber can also lead to an increment in the unburned combustion product.

## 4. Conclusions

The purpose of the present chapter is to demonstrate the influence of ethanol and methanol addition to gasoline on spark-ignition engine performance and emission characteristics. The summarized results from this study are the following:

With the increase of the percentage of ethanol in the blended fuel, the engine brake power decreased for various engine speeds.

With the increase of the percentage of methanol in the blends M5 and M10, the brake power slightly increased, and with the increase of the percentage of methanol in the blends M30 and M50, the brake power decreased.

As the ethanol (methanol) percentage increased, the BSFC increased. The blended fuels show higher BSFC and lower engine brake power than pure gasoline. Furthermore, there is a slight difference between the BSFC in comparison of gasoline and gasoline blended fuels (E5, E10, and E20 and M5, M10, and M20).

When there is an increase in ethanol and methanol percentage, the CO and HC concentration decreases. The lowest CO and HC emissions are obtained with blended fuel containing methanol (M50).

Increasing the percentage of ethanol and methanol leads to a significant increase in NOx emissions.

When there is an increase in the ethanol and methanol percentage up to 30% E30 (M30), there is an increase in the NOx concentration, followed by a decrease, after which it decreases with increasing ethanol (methanol) percentage. The lowest NOx emissions are obtained with gasoline.

## Acknowledgements

The present chapter has been written with the Project No 2018-RU-07's financial assistance. We are also eternally grateful to AVL-AST, Graz, Austria, for granting the use of AVL BOOST under the university partnership program.

## Author details

Simeon Iliev

Address all correspondence to: spi@uni-ruse.bg

Department of Engines and Vehicles, University of Ruse, Ruse, Bulgaria

## References

[1] Iliev S. Investigation of N-butanol blending with gasoline using a 1-d engine model. In: Special Session on Sustainable Mobility Solutions: Vehicle and Traffic Simulation, On-Road Trials and EV Charging. 2017. pp. 385-391. DOI: 10.5220/0006284703850391

[2] Melo T, Machado G, Machado RT, Pereira Belchior CR Jr, Pereira PP. Thermodynamic modeling of compression, combustion and expansion processes of gasoline, ethanol and natural gas with experimental validation on a flexible fuel engine. In: SAE World Congress. 2007. 2007-24-0035

[3] Varol Y, Oner C, Oztop HF, Altun S. Comparison of methanol, ethanol, or n-butanol blending with unleaded gasoline on exhaust emissions of an SI engine. Energy Sources Part A Recovery Utilization and Environmental Effects. 2014;**36**:938-948

[4] Pourkhesalian A, Shamekhi A, Salimi F. Alternative fuel and gasoline in an SI engine: A comparative study of performance and emissions characteristics. Fuel. 2010;**89**:1056-1063

[5] Chen C, Rao P, Delfino J. Oxygenated fuel induced to solvent effect on the dissolution of polynuclear aromatic hydrocarbons from contemned soil. Chemosphere. 2005;**60**:1572-1582

[6] Canaksi M, Ozsezen AN, Alptekin E. Impact of alcohol-gasoline fuel blends on exhaust emission on an SI engine. Renewable Energy. 2013;**52**:111-117

[7] Cavalcante Cordeiro de Melo T, Bastos Machado G, Machado RT, Pereira Belchior CR Jr, Pereira PP. Thermodynamic modeling of compression, combustion and expansion processes of gasoline, ethanol and natural gas with experimental validation on a flexible fuel engine. In: SAE World Congress. 2007. 2007-24-0035

[8] Pikunas A, Pukalskas S, Grabys J. Influence of composition of gasoline-ethanol blends on parameters of internal combustion engines. Journal of KONES Internal Combustion Engines. 2003;**10**:3-4

[9] Bata R, Elord A, Rice R. Emissions from IC engines fueled with alcohol–gasoline blends: A literature review. Journal of Engineering for Gas Turbines and Power. 1989;**111**:424-431

[10] Silva R, Cataluna R, Menezes EW, Samios D, Piatnicki CMS. Effect of additives on the antiknock properties and Reid vapor pressure of gasoline. Fuel. 2005;**84**:951-959

[11] Raveendran K, Ganesh A. Heating value of biomass and biomass pyrolysis products. Fuel. 1996;**75**:1715-1720

[12] Egebäck K, Henke M, Rehnlund B, Wallin M, Westerholm R. Blending of ethanol in gasoline for spark ignition engines—problem inventory and evaporative measurements. Rapport MTC 5407. 2005

[13] Palmer F. Vehicle performance of gasoline containing oxygenates. In: International Conference on Petroleum Based and Automotive Applications. London, UK: Institution of Mechanical Engineers Conference Publications, MEP; 1986. pp. 33-46

[14] Kim S, Dale BE. Global potential bioethanol production from wasted crops and crop residues. Biomass and Bioenergy. 2004;**26**(4):361-375

[15] Shenghua L, Clemente ERC, Tiegang H, Yanjv W. Study of spark ignition engine fueled with methanol/gasoline fuel blends. Applied Thermal Engineering. 2007;**27**:1904-1910

[16] Bilgin A, Sezer I. Effects of methanol addition to gasoline on the performance and fuel cost of a spark ignition engine. Energy & Fuels. 2008;**22**:2782-2788

[17] Altun S, Oztop H, Oner C, Varol Y. Exhaust emissions of methanol and ethanol-unleaded gasoline blends in a spark ignition engine. Thermal Science. 2013;**17**(1):291-297

[18] Shamekhi A, Khtibzade N, Shamekhi AH. Performance and emissions characteristics of a bi-fuel SI engine fueled by CNG and gasoline. ASME Paper. 2006. ICES2006-1387

[19] Onorati A, Montenegro G, D'Errico G. Prediction of the attenuation characteristics of IC engine silencers by 1-D and multi-D simulation models. SAE Technical Paper Series 2006. Tech. Rep. 2006-01-1541. 2006. DOI: 10.4271/2006-01-1541

[20] Montenegro G, Onorati A. Modeling of silencers for IC engine intake and exhaust systems by means of an integrated 1D-multiD approach. SAE International Journal of Engines. 2009;**1**(1):466. DOI: 10.4271/2008-01-0677

[21] Montenegro G, Onorati A, Piscaglia F, D'Errico G. Integrated 1D-multi-D fluid dynamic models for the simulation of ICE intake and exhaust systems. SAE Technical Paper Series 2007. Tech. Rep. 2007-01-0495. 2007. DOI: 10.4271/2007-01-0495

[22] Iliev S. Developing of a 1-D combustion model and study of engine performance and exhaust emissions using ethanol-gasoline blends. In: IAENG Transaction of Engineering Technologies. Netherlands: Springer; 2014, unpublished

[23] Iliev S. Developing of a 1-D combustion model and study of engine characteristics using ethanol-gasoline blends. In: Proceedings of the World Congress on Engineers. Vol. II. 2014, WCE 2014/ 978-988-19253-5-0

[24] AVL List Gmbh. AVL Boost—Theory. 2013

[25] Bowman C. Kinetics of pollutant formation and destruction in combustion. Progress in Energy and Combustion Science. 1975;**1**(1):33-45

[26] Onorati A, Ferrari G, D'Errico G. 1D unsteady flows with chemical reactions in the exhaust duct-system of S.I. engines: Predictions and experiments. SAE Paper No. 2001-01-0939

Chapter 10

# Biodiesel in Brazil Should Take Off with the Newly Introduced Domestic Biofuels Policy: RenovaBio

Fernando C. De Oliveira and Suani T. Coelho

Additional information is available at the end of the chapter

http://dx.doi.org/10.5772/intechopen.79670

**Abstract**

Our planet's climate has experienced changes mainly ascribed to the emission of carbon dioxide ($CO_2$), which accumulates in the atmosphere and causes an increase in the Earth's average temperature. In 2015, heads of state and scientists from several countries met in Paris to discuss measures aimed at curbing greenhouse gas emissions in order to limit that temperature rise to 2°C by the end of this century. As $CO_2$ needs to be banned from our environment for the sake of our own planet, it is reasonable for biofuels to present themselves as clean alternatives for the gradual replacement of fossil fuels. Biodiesel stands as an option. After 2005, some public policies were created in Brazil as an attempt to establish biodiesel as a replacement for mineral diesel, mainly in the transport sector. Although moderately successful, none of them compare in scope and reach to RenovaBio, a domestic biofuels policy that will make the production and use of biodiesel take off, once and for all. Therefore, herein, we present a brief overview on the status of the biodiesel production and use before the enactment of this new biofuels policy and the forecast of what it is expected to happen after its implementation.

**Keywords:** RenovaBio, biodiesel policy, fossil fuel, GHG, carbon dioxide, $CO_2$ emission

## 1. Introduction

The climate of our planet undergoes changes that mainly derive from the emission of greenhouse gases (GHGs), among which is carbon dioxide ($CO_2$), considered by many as the most significant man-made GHG. When expelled by the exhaust of internal combustion-powered vehicles, $CO_2$ accumulates in the atmosphere and causes an increase in the Earth's average temperature. As $CO_2$ has a very long atmospheric lifetime [1], usually longer than the other three main heat-trapping gases (methane, nitrous oxide, and CFCs—except for CFC-13 and CFC-115), it is seen as the chief culprit of global warming, which has worried the scientific community around the world in recent years.

Bearing in mind that it is essential to limit $CO_2$ emissions in the air, dozens of heads of state and hundreds of scientists from all over the globe gathered at the end of 2015, in Paris, for the 21st Conference of Parts—also known as COP21 or Paris Agreement—with the main goal of engaging nations in an international agreement on climate change in which everyone pledged to collaborate so that the increase in the average temperature of our planet should not exceed the limit of 2°C by the end of this century.

At the COP21, Brazil voluntarily committed to: (1) cut down on GHG emission up to 37% by 2025 and 43% by 2030—considering 2005 as the base year for both scenarios [2]—which would represent approximately 1.2 million tons of carbon dioxide equivalent ($CO_2$eq) by the year 2020 [3]; (2) increase the use of biofuels from sustainable bioenergy up to 18% of the total domestic energy mix by the year 2030 [4]; and (3) increase the share of renewables to an estimated 45% by the same year [5].

So, considering the scenario wherein $CO_2$ needs to be banned from our environment for the sake of our own planet, it is reasonable for biofuels to present themselves as clean and renewable alternatives [6] for the gradual replacement of fossil fuels such as gasoline and mineral diesel, for example. Therefore, because biodiesel is biodegradable, less toxic, and almost 100% renewable [7, 8], it stands as an option.

After 2005, some public policies were created in Brazil as an attempt to definitely establish biodiesel as replacement for fossil diesel, mainly in the transport sector. Although successful to some extent, none of them compare in scope and reach to the recently created National Biofuels Policy (RenovaBio), whereto there is a hope it will make the production and use of biodiesel increase at a much faster rate.

Therefore, this work is primarily aimed at informing the reader about the past biodiesel policies that were implemented to boost its production and use, providing special emphasis to the most recent one: RenovaBio.

## 2. Early stages of biodiesel in Brazil and first public policies

The first time Brazil began to flirt with biodiesel was during the 1920s, when the *Instituto Nacional de Tecnologia*, a government technological institution, began to undertake studies on that biofuel [9].

Although the first real tests with biodiesel in Brazil began to appear 20 years later, in the early 1940s, during World War II [10], it was only during the 1970s, after continued increases in oil prices, that conventional biodiesel technology, using a catalyst and an alcohol, became well known in Brazil [11].

With the oil shock of 1973, the world awakened to the importance of finding different sources of energy that were not dirty like those produced by the fossil fuels. That year became a landmark in our planet's energy history, whose focus would be on overcoming the energy crisis via two main groups of action: conservation of energy—or energy saving—and the use of alternative sources of energy [12].

Following that type of awareness, Expedito Parente, the leading Brazilian biodiesel scientist at that time, created the first patents of that biofuel in the country, which would represent the outcome of his studies in the late 1970s and early 1980s. As a result, he became the mainstay and principal proponent of the early developments of biodiesel in Brazil.

The first policies created to promote the production and consumption of biodiesel in Brazil (**Table 1**) are discussed ahead.

### 2.1. Social Fuel Stamp (SCS)

The Brazilian government launched the Social Fuel Stamp program—or *Selo Combustível Social (SCS)*, in Portuguese—in December 2004, through Decree No. 5297 [13], to promote social inclusion of family farmers from the two poorest regions of Brazil: North and Northeast [14]. The intention was also to regionalize the production—highly dependent on soybean and beef tallow—of other abundant raw materials, such as palm, cotton seed, and maybe jatropha [15].

Plentiful in both regions, castor would be a nice contender if it were not for the fact that this oilseed presents a very high viscosity content of 14.1 $mm^2/s$, which is way above the maximum determined by the Brazilian Petroleum Agency (ANP) of 6.0 $mm^2/s$ [16]. That technical constraint alone deems castor an unsuitable alternative feedstock for the production of biodiesel [10, 17–20].

The SCS is also a mechanism that serves to reduce government taxes and allow the Brazilian Development Bank (BNDES) to grant lower funding fees for biodiesel producers who acquire raw materials from—and provide technical assistance to—family farmers [15].

Through ANP, the Federal government promotes auctions for the purchase of biodiesel. During the first stage of the auction, 80% of the total volume is offered to biodiesel producers who already have the SCS. This phase, restricted to holders of the stamp, grants them the purchase and delivery of the product for specified periods. The remaining 20% of biodiesel is offered to any producer, with or without the stamp [21, 22].

| Year | Mechanism | Program name and acronym (in Portuguese) |
|---|---|---|
| 2004 | Decree No. 5297 | Social Fuel Stamp (SCS) |
| 2005 | Law No. 11097 | National Program of Production and Use of Biodiesel (PNPB) |
| 2009 | Law No. 12187 | National Policy on Climate Change (PNMC) |
| 2014 | Law No. 13033 | Mandatory blend on diesel: increase to 6% and 7% |
| 2016 | Law No. 13263 | Mandatory blend on diesel: increase to 8%, 9% and 10% |
| 2017 | Law No. 13576 | National Biofuels Policy (RenovaBio) |

**Table 1.** Policies that helped biodiesel in Brazil.

Although the SCS program was conceived to bring about major social and economic changes in the North and Northeast regions, as initially intended by the government, except for some modest progress, many authors see it as a failure. The reasons vary from high logistic and harvesting costs to low economies of scale, low utilization of agricultural machinery, artificial irrigation and fertilizer, and lack of access to financing due to red tape [18]. Therefore, this program is currently under revision in order to correct the aforesaid problems, already realized by the government.

### 2.2. National Program of Production and Use of Biodiesel (PNPB)

On January 13, 2005, the Brazilian government formally introduced The National Program of Production and Use of Biodiesel (PNPB, Brazilian acronym for *Programa Nacional de Produção e Uso de Biodiesel*) into the energy matrix through the implementation of Federal Law No. 11097 [23].

One of the main objectives of PNPB, besides the promotion of social inclusion of family farmers, primarily from the North and Northeast regions of Brazil, was also to reduce the import of mineral diesel [9, 24], thus positively affecting the Brazilian trade balance.

In addition, the program also meant to establish a minimum blending percentage of biodiesel into petroleum diesel. At first, the suggested blending percentage (2%) was optional to run from 2005 to 2007, then mandatory from January 2008 onward (**Table 2**).

**Table 2** shows that the production percentage rates from 2005 to 2010 were (on average) very high when compared to the ones from 2011 to 2017. Although 2014 was a year in which the blend changed twice, the percentage increase was relatively low: only 17%. And since then, the percentage has declined, even negative in 2016—a year in which there was a decrease in the biodiesel production for the first time since the establishment of the PNPB program, in 2005.

Right from the beginning of the PNPB program, soybean has become the main and overwhelmingly dominant raw material for biodiesel production, accounting for an average of 75% and beef tallow, in second, with an average of 15%. Other raw materials, such as cotton seed, used cooking oil, and other fats would represent, altogether, the remaining 10% [25]. This high dominance of both feedstocks has driven the Brazilian government to create mechanisms to incentivize the use of alternative raw materials [15, 26].

### 2.3. National Policy on Climate Change (PNMC)

Law No. 12187 was sanctioned on December 29, 2009, and amended by the Presidential Decree No. 7390 on December 9, 2010 [27]. Together, they establish principles, goals, instruments, and guidelines of the National Policy on Climate Change (PNMC) and elucidate terms such as adaptation, mitigation, emissions and sources of emissions, greenhouse gases, and their eminent impacts [27].

Both law and decree also provide that any measures taken to reduce any type of emissions must have a national range, besides being focused on the prevention or minimization of damages caused directly by anthropogenic activities. On the other hand, these measures should also consider the different socioeconomic contexts of their application, as well as allocate to the population and economic sectors the burdens and charges resulting from such impacts.

| Year | B100 ($m^3$) | % | Blend |
|---|---|---|---|
| 2005 | 736 | — | B2 |
| 2006 | 69,002 | 9275 | B2 |
| 2007 | 404,329 | 486 | B2 |
| 2008 | 1,167,128 | 189 | B2/B3 |
| 2009 | 1,608,448 | 38 | B3/B4 |
| 2010 | 2,386,399 | 48 | B5 |
| 2011 | 2,672,760 | 12 | B5 |
| 2012 | 2,717,483 | 2 | B5 |
| 2013 | 2,917,488 | 7 | B5 |
| 2014 | 3,422,210 | 17 | B5/B6/B7 |
| 2015 | 3,937,269 | 15 | B7 |
| 2016 | 3,801,339 | −3 | B7 |
| 2017 | 4,291,294 | 13 | B8 |
| 2018 | 5,590,000* | 30 | B10 |

*Expected demand [21].

**Table 2.** Annual biodiesel production and respective rounded percentages [25].

Furthermore, both policies also shed light on how to diminish natural and anthropogenic impacts, and how to promote the understanding about the consequences of existing and forthcoming climate change events [27].

One of PNMC's guidelines is Brazil's Nationally Determined Contributions (NDC) committed at the Paris Agreement.

## 2.4. Blending mandates on diesel: 6 and 7%

On September 24, 2014, the Federal government enacted Law No. 13033 [28], turning mandatory the blending of biodiesel into petroleum diesel in the following percentages: 6% (B6) beginning on July 1, 2014, and 7% (B7) starting on November 1 of the same year.

This policy also allows the voluntary addition of biodiesel to diesel in amounts greater than the mandatory percentage in public transport, rail, inland navigation, equipment, or vehicles for mineral extraction and electric power generation, tractors, and other automotive vehicles intended to pull or tow agricultural machinery or to carry out agricultural work [28].

## 2.5. Blending mandates on diesel: 8 and 10%

Law No. 13263, implemented on March 23, 2016, alters Law No. 13033 to establish new mandatory blending percentages of biodiesel into mineral diesel, as follows: 8, 9, and 10 (B8, B9, B10) to commence on March 1 of 2017, 2018, and 2019, respectively [29]. But the latter policy upholds the prerogative of voluntary addition of the former.

This law also authorizes the increment of up to 15% (B15) of biodiesel to fossil diesel sold to end users, in any part of the country, after completion (up to 36 months of its enactment) of specific tests and experiments on engines that validate the use of the mixture [29]. The specific tests (50 or so) shall be conducted by more than 20 different companies, spread out through more than 15 locations nationwide [21].

With the anticipation of the B10 blend to March 2018, there should be a progressive increase in the biodiesel-to-diesel mix aiming to achieve the B15 in 2025 in order to meet the country's NDC, committed at the Paris Agreement [21].

## 3. The most recent public policy on biofuels: RenovaBio

### 3.1. What is RenovaBio?

The National Policy on Biofuels or *Política Nacional de Biocombustíveis* (RenovaBio) is an integral part of the national energy policy to consolidate the production and use of biofuels in Brazil. This program, converted into Law No. 13576 by the Federal government on December 26, 2017 [30], aims to radically expand the production and use of biodiesel, biomethane, and ethanol, among other biofuels, and increase energy security throughout the country and, therefore, create more jobs and income, promote a greater social insertion of family farmers, and help Brazil meet its GHG emission reduction targets, committed at the COP21 [30].

Besides helping the biodiesel market, RenovaBio will also boost the ethanol industry and Brazil's foreign trade balance by diminishing the country's reliance on imported corn ethanol from the United States.

With a nationwide scope, this program also outlines rules for marketing biofuels in the country under the flag of environmental sustainability, and fosters credibility and predictability of national fuel supply. By doing so, RenovaBio will not just help the environment, but it will lay the market conditions for domestic and foreign private investments [31, 32].

This fresh and innovative regulatory framework is supported by two main pillars: the encouragement of energy efficiency throughout the production and use of biofuels and the recognition of biofuels capacity to remove carbon from the environment [33], or mitigate its impacts.

When formulating the basic tenets of RenovaBio, policy-makers and researchers took into account successful international models of initiatives that were enacted for the areas of biofuels and renewable energy, such as the Low Carbon Fuel Standard (LCFS), the Renewable Fuel Standard (RFS), and the Renewable Energy Directive (RED) [34, 35].

Both LCFS and RFS laws, implemented in the United States in 2005 and 2007, respectively, are primarily aimed at reducing carbon intensity in the transport sector [36, 37], while the RED directive, sanctioned in the Europe Union, in 2009, demand that each EU country expressively increase their share of renewable energy to the energy mix in a way that the overall EU share be 20% by the year 2020 [38].

http://dx.doi.org/10.5772/intechopen.79670

### 3.2. RenovaBio and Brazil's commitments to mitigate carbon dioxide emissions

Biofuels have aroused the interest of the Brazilian government in the light of its mitigating potential against the harm caused by petroleum-derived fuels, mainly in the transport sector [22]. Therefore, by enacting RenovaBio, the Brazilian government has overtly shown to the international scientific community its desire to comply with the Paris Agreement, in which the country presented its voluntary goals for 2030 under the NDC.

To achieve these goals, Brazil will adopt actions to reduce GHG projected emissions and increase the share of renewable energy and biofuels into the country's energy matrix [39], as aforementioned.

Furthermore, there is a possibility of voluntary use of biodiesel in mineral diesel to be greater than the mandatory blend in specific cases, such as: 20% in captive fleets or road users serviced by supply point; 30% in rail transport; 30% in agricultural and industrial use; 100% in experimental use, specific or in other applications [21].

Decree No. 9308, sanctioned on March 15, 2018, addresses the annual compulsory targets to reduce domestic emissions of GHGs and assigns to the National Energy Policy Council (CNPE) the definition of these goals, which are based on recommendations of the Interministerial Committee on Climate Change (CIM) [40].

The breaking down of the national mandatory targets—provided by CNPE—into an individual goal that must be assigned to each biofuel distributor is responsibility of ANP. The distributors that do not comply with their individual goals will be subject to a fine proportional to the noncompliance goal, which may not exceed 5% of their annual revenue, registered in the previous two years [40].

This decree has also established a RenovaBio Committee to provide technical support to CNPE in the process of defining—until June 2018—the annual national reduction targets and it is consisted of agents from seven Ministries, among them Mines and Energy, Environment, and Agriculture. Representatives of other federal, state, and municipal agencies, as well as public and private sector entities from the biofuels market, in addition to technicians and specialists from the sector, whose assistance will be considered provision of unpaid public services, may also be part of this committee as guests [40].

One of the new facets of RenovaBio, missing in previous biofuels policies, is the creation of two important market mechanisms: *Certificado de Produção Eficiente de Biocombustíveis* (CPEB), or Certificate of Efficient Production of Biofuels, and *Créditos de Descarbonização* (CBios), or Decarbonization Credits.

#### *3.2.1. Certificate of Efficient Production of Biofuels (CPEB)*

Chapter 2 of the RenovaBio policy defines the CPEB as a document issued exclusively by inspection companies as a result of the biofuel certification process. These companies must be highly qualified as such (ISO-standard) and be approved by the government in order to inspect the biofuel companies—either producers or importers [30].

After the inspection and certification are over, ANP will audit the whole process in order to approve or disapprove the issuance of the certificate. In case of approval, the list of certified producers and/or importers is published in the official government gazette so they can take advantage of their certification status when marketing their biofuels.

In case of disapproval, ANP then assesses the whole process to see if there was incompleteness or any sort of fraud. If the latter is the case, then ANP establishes an administrative process to revoke the accreditation of the inspection company. If the former is the problem, then ANP sends the whole paperwork to the inspection company redo the process again [41], as shown in **Figure 1**.

Article 28 of the RenovaBio law states that a bonus of up to 20% will be applied on the value of the energy and environmental efficiency grade of the producer or importer of biofuel whose CPEB proves a negative emission of GHGs in its life cycle in relation to its substitute of fossil origin.

It is up to the RenovaBio Committee, among other responsibilities, to monitor the market, supply, and development of the production of biofuels, particularly the installed capacity of companies that hold the certificate that deems their production efficient and environmentally friendly. Government Order No. 103 of March 22, 2018, has set the green light for this Committee to start its activities immediately [42].

### *3.2.2. Decarbonization Credits (CBios)*

Chapter 2 of the RenovaBio law states that Decarbonization Credits are instruments registered in the form of scripture for the purpose of attesting the individual target of fuel distributors whose proof of achievement shall be based on the amount of credits held by the fuel distributor on the date defined by the policy [30].

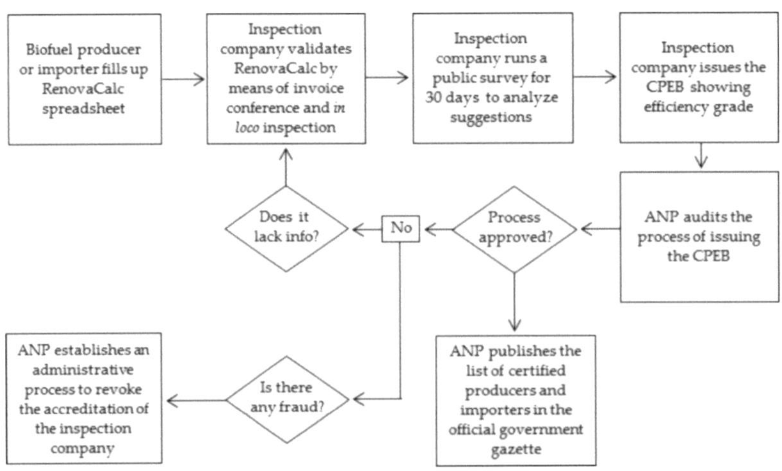


**Figure 1.** CPEB issuance process flowchart. Source: Adapted from [41].

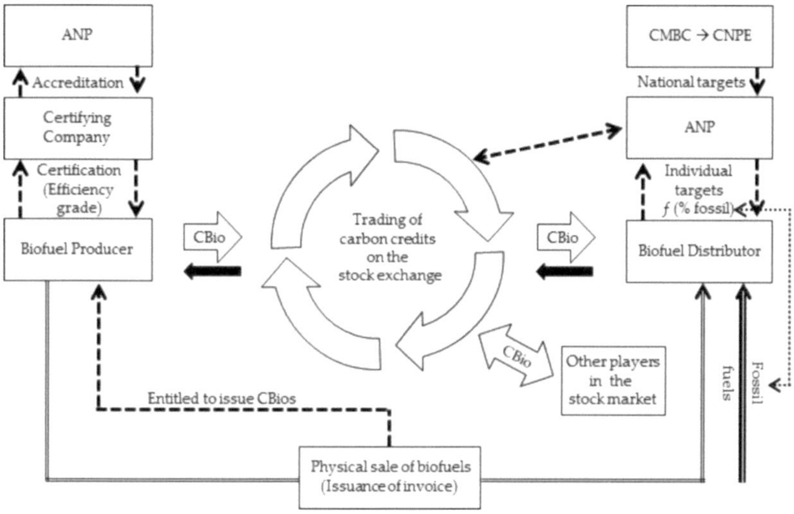


**Figure 2.** CBios flowchart. Source: Adapted from [41].

The values of the annual mandatory targets were established in units of CBios, with each CBio corresponding to 1 ton $CO_2eq$, whose calculations consider the difference between the GHG emissions in the life cycle of a biofuel and the emissions in the life cycle of its fossil substitute [40].

These calculations will consider energy efficiency in MJ/ton or cbm and environmental impact in $gCO_2e$/ton or cbm [33] and will be carried out by RenovaCalc, which is a tool that accounts for the carbon intensity of a biofuel in $gCO_2eq/MJ$ [43].

Besides carrying out analyses and studies for the definition of compulsory goals, plus the evaluation and suggestion of preventive measures to adapt them, other responsibilities of the RenovaBio Committee are the monitoring of supply, demand, and price of CBios issued and traded in the stock exchange from the commercialization of biofuels. **Figure 2** shows the links of CBios with the biofuel producers and distributors and how they interplay with the government agencies and certifying companies.

## 4. Current and future production, consumption, and capacity

### 4.1. Production, consumption, and installed capacity before RenovaBio

Historically, Brazil has adopted soybean and beef tallow as the main feedstocks for the fulfillment of biodiesel demand in the country. As previously mentioned, soybean alone accounts for three-fourths of the domestic production, mostly done in the Mid-West and in the South, regions that house the vast majority of Brazil's biodiesel plants, whose owners are also major soy producers and agribusiness companies, well established in both areas for a long time and, therefore, are better prepared to take advantage of the big soy market that was developed throughout the country [3].

Since the inception of the PNPB program, in 2005, biodiesel production has leap-frogged from 736 thousand to 4.3 billion liters in just 12 years (**Figure 3**). Such a growth turned Brazil into the second biodiesel producer in the world, trailing only the United States, as the number one producer with approximately 5.5 billion liters, in 2016 [44].

In footstep with the production growth, the installed capacity also jumped from zero to approximately 8 billion liters in about the same period (**Figure 3**). If on one hand, this growth leads to an idle capacity rate that is close to 50%; on the other hand, it makes biodiesel producers optimists to comfortably meet the production forecasts for at least 5 years after the enactment of the new biofuel policy.

### 4.2. Production, consumption, and installed capacity after RenovaBio (forecast)

The expected biodiesel production growth to 18 billion liters in 2030 suggests a major boost of the installed capacity to around 22 billion liters. In order to meet that forecast, it will be

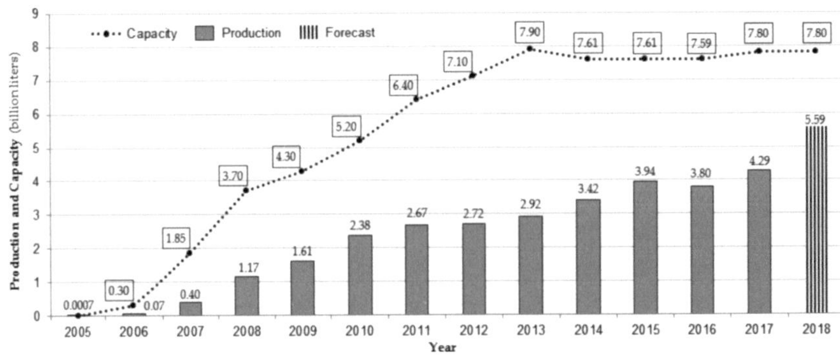


**Figure 3.** Biodiesel annual production (B100) and accumulated installed capacity. Source: Adapted from [25].

| Status | Accomplished | Forecasted | | | Unit (million) |
|---|---|---|---|---|---|
| Year | 2016 | 2020 | 2025 | 2030 | |
| Soybean | | | | | |
| Processed | 40.7 | 55.1 | 77.3 | 107.2 | t/year |
| Installed capacity | 65.0 | 68.8 | 96.6 | 134.0 | t/year |
| Number of plants | 117 | 120 | 139 | 165 | — |
| Biodiesel | | | | | |
| Production | 3.8 | 6.4 | 11.4 | 18.0 | $m^3$/year |
| Installed capacity | 7.3 | 8.0 | 14.3 | 22.5 | $m^3$/year |
| Number of plants | 50 | 51 | 76 | 109 | — |

Source: Adapted from [45].

**Table 3.** Future scenario for soybean processing units and biodiesel refining.

necessary not just to expand the capacity of current plants but also a twofold increase in the number of biodiesel plants in operation now (**Table 3**).

That kind of growth will require an investment volume of R$ 21.7 billion [45], equivalent to approximately US$ 7 billion, from the government and the private sector, which will represent a major boost on the local economy where the current biodiesel plants are already installed, as well as where the new ones should be built.

Such expansion in the production and use of biodiesel should represent some positive externalities like the creation of new jobs along the entire biodiesel chain, and the reduction of $CO_2$ emissions in the atmosphere as a replacement for part of the petroleum diesel Brazil still needs to import, which should represent an economy of US$ 1 billion per year for an equivalent amount of nearly 1.2 billion liters of diesel not imported [3].

## 5. What still needs to be done in spite of RenovaBio

The complexity of this policy requires the government to pay special attention to questions on how the process steps will be supervised and how to allocate the individual targets of the biofuel distributors which, in turn, can acquire the decarbonization certificates whose prices still need to be defined, as well as the costs for the certification of production [46].

The technology mostly used to produce biodiesel in Brazil is the transesterification, which is inefficient. Besides, the process uses methanol, which is from a fossil source, instead of ethanol, which is cleaner, renewable, and produced locally from sugarcane. And the country still imports a good chunk of the alcohol that is used to produce biodiesel, therefore affecting the trade balance negatively.

Furthermore, there needs to be greater incentives for the diversification of the mix of raw materials used in the production of biodiesel, since soybeans and beef tallow together represent about 90% of the total and they present environmental problems due to the use of pesticides and herbicides [47], as well as GHG from land use and land use change [48].

Perhaps alternative sources, such as palm—whose yield per hectare is approximately six times greater than that of soybeans [22]—may be an option. However, this feedstock still needs investments in R&D so that its production increases in such a way that the amount of available oil would be enough not only to meet its main market—culinary—but also the production of biodiesel.

Although at this moment in time palm does not have enough scale to meet the market demand for biodiesel, Embrapa's research with this oilseed has had positive results in adapting it to other environments that are different from the traditional ones in the legal Amazon, where most of palm is harvested [21].

Thus, it is hoped that palm oil will have a much larger penetration in the mix of raw materials for the production of biodiesel and, therefore, provide a greater competitiveness of the sector and increase the income of family farmers, especially those from the north and northeast. But that will depend as well on the success of policies such as *Propalma—Programa de Produção Sustentável da Palma de Óleo no Brasil*, or the Brazilian Program of Sustainable Palm Oil Production [21].

On the overseas front, the European Union has recently strengthened its position to disfavoring traditional biofuels, such as ethanol from sugarcane and corn, and biodiesel from oilseeds. The maximum demand for these biofuels in energy demand is forecasted to decrease from 7% in 2020 to 3.8% in 2030. This proposal has received harsh criticisms from various sectors of the industry [21], as well as from the scientific community.

At a moment in which the demand for food and bioenergy will continue to grow for the next years as a result of populational growth, increased world income, and the evolution of energy policies toward cleaner and more sustainable sources [21], the European Union position, along with the United States decision to withdraw from the Paris Agreement, raises concerns regarding the immediate growth of biofuels worldwide. On the other hand, Brazil has now a great opportunity to set the stage to confirm its world leadership in the field of biofuels [33].

## 6. Conclusion

Over the last years, we have noticed that a clear global movement is underway to reduce the use of fossil fuels in order to decrease the GHG emission in the atmosphere and, therefore, limit the rise of the temperature of our planet to 2°C, preferably to 1.5. One of the ways to achieve this reduction is by replacing fossil fuels used in motor vehicles with renewable ones such as biodiesel.

The enactment of the latest and most innovative domestic biofuels policy to date, RenovaBio, establishes a legal framework for its production and use in the country by promoting energy efficiency and competitiveness of the sector through meritocracy.

By allowing greater market predictability for the entire biofuels chain, RenovaBio poises to take this industry in Brazil to new heights by creating more jobs and boosting the economy with major investments from the government and private sectors.

However, there are some problems that may undermine the whole program if not addressed accordingly. For example, at the time of the creation of the Social Fuel Stamp, the focus was put on the social side of the program, neglecting the technical aspects that deemed that initiative a failure, as in the case of castor and palm feedstocks.

Another major problem lies in the technology used to produce biodiesel in Brazil: the transesterification, which is inefficient. Also, the process uses methanol—which besides from being a fossil source, the country still imports a good chunk of it—instead of ethanol, which is cleaner, renewable, and produced from the local sugarcane.

Furthermore, the main raw materials used to produce biodiesel in Brazil (soy and beef tallow) present some environmental problems that are already known by the government and industry experts, largely due to the use of pesticides and herbicides, in addition to GHG from land use and land use change.

## Acknowledgements

This work was supported by grants from CAPES and FAPESP through grant number 2012/51466-7 and grant number 2014/50279-4.

## Author details

Fernando C. De Oliveira* and Suani T. Coelho

*Address all correspondence to: folive@usp.br

University of Sao Paulo, Institute of Energy and Environment, Bioenergy Research Group, São Paulo, Brazil

## References

[1] IPCC—Intergovernmental Panel on Climate Change. IPCC Fourth Assessment Report: Climate Change 2007 [Internet]. 2007. Available from: http://www.ipcc.ch/publications_and_data/ar4/wg1/en/ch2s2-10-2.html#table-2-14 [Accessed: May 04, 2018]

[2] Oliveira MIL. RenovaBio: por uma nova política nacional de biocombustíveis. Revista Opiniões. 2017;**54**:6-7. ISSN 2177-6504. Available from: http://revistaonline.revistaopinioes.com.br/revistas/suc/155/#page/6 [Accessed: May 16, 2018]

[3] De Oliveira FC, Coelho ST. History, evolution, and environmental impact of biodiesel in Brazil: A review. Renewable and Sustainable Energy Reviews. 2017;**75**:168-179. DOI: 10.1016/j.rser.2016.10.060

[4] Moreno-Perez OM, Marcossi GPC, Ortiz-Miranda D. Taking stock of the evolution of the biodiesel industry in Brazil: Business concentration and structural traits. Energy Policy. 2017;**110**:525-533. DOI: 10.1016/j.enpol.2017.09.001

[5] Tolmasquim M, Gorini R, Matsumura E, Soares J, Oliveira L, et al. The Brazilian Commitment to Combating Climate Change: Energy Production and Use. Rio de Janeiro: EPE; 2016. p. 96

[6] Araújo CDM, Andrade CCA, Silva ES, Dupas FA. Biodiesel production from used cooking oil: A review. Renewable and Sustainable Energy Reviews. 2013;**27**:445-452. DOI: 10.1016/j.rser.2013.06.014

[7] Barufi C, Pavan M, Zanotti JR, Soares M. Biodiesel e os dilemas da inclusão social. In: Bermann C, editor. As novas energias no Brasil. Rio de Janeiro: Fase; 2007. pp. 19-86

[8] Botelho CAV. Viabilidade técnica e aspectos ambientais do biodiesel etílico de óleos residuais de fritura [dissertation]. São Paulo: Universidade de São Paulo; 2012

[9] Bailis R. Brazil: Biodiesel. In: Solomon B, Bailis R, editors. Sustainable Development of Biofuels in Latin America and the Caribbean. New York: Springer; 2014. pp. 103-126

[10] Rico JAP, Sauer IL. A review of Brazilian biodiesel experiences. Renewable and Sustainable Energy Reviews. 2015;**45**:513-529. DOI: 10.1016/j.rser.2015.01.028

[11] Nogueira LAH. Does biodiesel make sense? Energy. 2011;**36**:3659-3666. DOI: 10.1016/j.energy.2010.08.035

[12] Parente EJS. Biodiesel: Uma Aventura Tecnológica num País Engraçado. Vol. 66. Fortaleza: Tecbio; 2003

[13] PLANALTO. Decree No. 5297—Criação do Selo Combustível Social (SCS) [Internet]. 2004. Available from: http://www.planalto.gov.br/ccivil_03/_ato2004-2006/2005/Lei/L11097 [Accessed: May 05, 2018]

[14] MDA—Ministério do Desenvolvimento Agrário. O Selo Combustível Social. SAFn.d [Internet]. 2004. Available from: http://www.mda.gov.br/sitemda/secretaria/saf-biodiesel/o-selo-combust%C3%ADvel-social [Accessed: May 11, 2018]

[15] IPEA—Instituto de Pesquisa Econômica Aplicada. Biodiesel no Brasil: desafios das políticas públicas para a dinamização da produção. Comunicado Nr. 137 [Internet]. 2012. Available from: http://www.ipea.gov.br/portal/index.php?option=com_content&view=article&id=13354 [Accessed: May 05, 2018]

[16] Rathmann R, Szklo A, Schaeffer R. Targets and results of the Brazilian Biodiesel Incentive Program—Has it reached the Promised Land? Applied Energy. 2012;**97**:91-100. DOI: 10.1016/j.apenergy.2011.11.021

[17] Santana GCS, Martins PF, Silva NL, Batistella CB, Maciel Filho R, et al. Simulation and cost estimate for biodiesel production using castor oil. Chemical Engineering Research and Design. 2010;**88**:626-632. DOI: 10.1016/j.cherd.2009.09.015

[18] Kamimura A, Oliveira AN, Burani GF. Brazilian family farming agriculture in the biodiesel production: A portrait of regional possibilities. Low Carbon Economy. 2011;**2**:7-14. DOI: 10.4236/lce.2011.21002

[19] Angarita EE, Rocha MH, Lora EES, Venturini OJ, Torres EA, et al. Biocombustíveis de primeira geração: biodiesel. In: Lora EES, Venturini OJ, editors. Biocombustíveis. 1st ed. Vol. 1. Rio de Janeiro: Interciência; 2012. pp. 173-309

[20] Suarez P. Crônica de uma morte anunciada. o óleo de mamona como matéria-prima para produção de biodiesel. BiodieselBR. 2015;**44**:46

[21] EPE—Empresa de Pesquisa Energética. Plano Decenal de Expansão de Energia 2026 [Internet]. 2017. Available from: http://www.epe.gov.br/pt/publicacoes-dados-abertos/publicacoes/Plano-Decenal-de-Expansao-de-Energia-2026 [Accessed: May 12, 2018]

[22] EPE—Empresa de Pesquisa Energética. RenovaBio—Biocombustíveis 2030: Papel dos biocombustíveis na matriz [Internet]. 2017. Available from: http://www.epe.gov.br/pt/publicacoes-dados-abertos/publicacoes/renovabio [Accessed: May 12, 2018]

[23] PLANALTO. Law No. 11097—Introdução do Biodiesel na Matriz Energética Brasileira [Internet]. 2005. Available from: http://www.planalto.gov.br/ccivil_03/_ato2004-2006/2005/Lei/L11097 [Accessed: May 05, 2018]

[24] Padula AD, Santos MS, Ferreira L, Borenstein D. The emergence of the biodiesel industry in Brazil: Current figures and future prospects. Energy Policy. 2012;**44**:395-405. DOI: 10.1016/j.enpol.2012.02.003

[25] ANP—Agência Nacional do Petróleo. Boletim Mensal do Biodiesel. Produção Nacional de Biodiesel Puro B100 (m3) [Internet]. 2018. Available from: http://www.anp.gov.br/wwwanp/dados-estatisticos [Accessed: October 02, 2018]

[26] Vaccaro GLR, Pohlmann C, Lima AC, Dos Santos MS, De Souza CB, et al. Prospective scenarios for the biodiesel chain of a Brazilian state. Renewable and Sustainable Energy Reviews. 2010;**14**:1263-1272. DOI: 10.1016/j.rser.2009.12.008

[27] PLANALTO. Law No. 12187—Política Nacional sobre Mudança do Clima [Internet]. 2009. Available from: http://www.planalto.gov.br/ccivil_03/_ato2007-2010/2009/lei/l12187.htm [Accessed: May 05, 2018]

[28] PLANALTO. Law No. 13033—Adição Obrigatória de Biodiesel ao Óleo Diesel [Internet]. 2014. Available from: http://www.planalto.gov.br/ccivil_03/_Ato2011-2014/2014/Lei/L13033 [Accessed: May 05, 2018]

[29] PLANALTO. Law No. 13263—Adição Obrigatória de Biodiesel ao Óleo Diesel [Internet]. 2016. Available from: http://www.planalto.gov.br/ccivil_03/_Ato2015-2018/2016/Lei/L13263.htm#art1 [Accessed: May 05, 2018]

[30] PLANALTO. Law No. 13576—Política Nacional de Biocombustíveis (RenovaBio) [Internet]. 2017. Available from: http://www.planalto.gov.br/ccivil_03/_Ato2015-2018/2017/Lei/L13576.htm [Accessed: May 05, 2018]

[31] Nastari P. Historical Breakthrough for Biofuels. Column about Sugar and Ethanol [Internet]. 2018. Available from: http://portal.datagro.com/en/sugar-etanol/2/plinio-nastari/82297/historical-breakthrough-for-biofuels [Accessed: May 17, 2018]

[32] Voegele E. Brazil's President Signs National Biofuels Policy into Law. Biodiesel Magazine January 3, 2018 [Internet]. 2018. Available from: http://www.biodieselmagazine.com/articles/2516247/brazils-president-signs-national-biofuels-policy-into-law [Accessed: May 16, 2018]

[33] Nastari P. Overcoming current challenges and framing the policy options ahead: Brazil's new proposal for a biofuels national policy RenovaBio. In: Proceedings of the Biofuture Summit 17; 24-25 October 2017; São Paulo. Brasilia: Biofuture Platform; 2017

[34] Addington B. RenovaBio: A Paradigm Shift for Biofuels in Brazil [Internet]. 2018. Available from: http://info.opisnet.com/renovabio-a-paradigm-shift-for-biofuels-in-brazil or from: http://www.udop.com.br/download/noticias/2017/24_07_17_artigo_renovabio.pdf [Accessed: May 16, 2018]

[35] Nastari P. A eletrificação com combustíveis e o RenovaBio (booklet in pdf) [Internet]. 2018. Available from: http://www.datagro.com [Accessed: May 17, 2018]

[36] AFDC—Alternative Fuel Data Center. U. S. Department of Energy. Renewable Fuel Standard [Internet]. 2018. Available from: https://www.afdc.energy.gov/laws/RFS [Accessed: May 16, 2018]

[37] CARB—California Air Resources Board. Low Carbon Fuel Standard [Internet]. 2018. Available from: https://www.arb.ca.gov/fuels/lcfs/lcfs.htm [Accessed: May 16, 2018]

[38] EURACTIV. Deal Secured on Ambitious EU Renewables Law [Internet]. 2008. Available from: https://www.euractiv.com/section/energy/news/deal-secured-on-ambitious-eu-renewables-law/ [Accessed: May 16, 2018]

[39] EPE—Empresa de Pesquisa Energética. The Brazilian Commitment to Combating Climate Change: Energy Production and Use. Rio de Janeiro: EPE, 2016. 96 p. Available from: www.epe.gov.br [Accessed: May 16, 2018]

[40] PLANALTO. Decree No. 9308—Definição de Metas Compulsórias Anuais de Redução de Emissões de Gases de Efeito Estufa [Internet]. 2018. Available from: http://www.planalto.gov.br/ccivil_03/_Ato2015-2018/2018/Decreto/D9308.htm [Accessed: May 16, 2018]

[41] Amaral ACN. Regulação do RenovaBio. In: Proceedings of the RenovaBio Próximos Passos (Speech); 1 February 2018; Rio de Janeiro. Rio de Janeiro: Centro Cultural FGV, 2018. Available from: http://www.anp.gov.br/palestra/4309-regulacao-do-renovabio (PowerPoint presentation) [Accessed: May 21, 2018]

[42] MME—Ministério das Minas e Energia. RenovaBio [Internet]. 2018. Available from: http://www.mme.gov.br/web/guest/pagina-inicial/outras-noticas/-/asset_publisher/32hLrOzMKwWb/content/portaria-da-inicio-as-atividades-do-comite-renovabio;jsessionid=65F286C75031E63B7DB6D2226A795B12.srv155?redirect=http%3A%2F%2Fwww.mme.gov.br%2Fweb%2Fguest%2Fpagina-inicial%2Foutras-noticas%3Bjsessionid%3D65F286C75031E63B7DB6D2226A795B12.srv155%3Fp_p_id%3D101_INSTANCE_32hLrOzMKwWb%26p_p_lifecycle%3D0%26p_p_state%3Dnormal%26p_p_mode%3Dview%26p_p_col_id%3Dcolumn-1%26p_p_col_pos%3D1%26p_p_col_count%3D3 [Accessed: May 11, 2018]

[43] Matsuura MI, Scachetti MT, Chagas MF, Seabra JE, Moreira MM, et al. RenovaCalc: Método e ferramenta para a contabilidade da intensidade de carbono e biocombustíveis no programa RenovaBio [Internet]. 2018. Available from: http://www.anp.gov.br/consultas-audiencias-publicas/em-andamento/4469-consulta-e-audiencia-publicas-n-10-2018 [Accessed: May 21, 2018]

[44] REN21—Renewable Energy Policy Network 21st Century. Renewables 2017 Global Status Report [Internet]. 2017. Available from: http://www.ren21.net/status-of-renewables/global-status-report [Accessed: May 16, 2018]

[45] Trigueirinho F, Minelli JC, Tokarski D. Biodiesel: oportunidades e desafios no longo prazo [Internet]. 2016. Available from: http://www.abiove.org.br/site/_FILES/Portugues/07102016-131231-07_10_2016_n-_cenario_para_o_biodiesel_em_2030(2).pdf [Accessed: May 11, 2018]

[46] Roitman T. RenovaBio—Próximos passos. Caderno de Biocombustíveis e o RenovaBio [Internet]. 2018. Available from: http://fgvenergia.fgv.br/eventos [Accessed: May 16, 2018]

[47] Cavalett O, Ortega E. Integrated environmental assessment of biodiesel production from soybean in Brazil. Journal of Cleaner Production. 2010;**18**:55-70. DOI: 10.1016/j.jclepro.2009.09.008

[48] Castanheira ÉG, Grisoli R, Freire F, Pecora V, Coelho ST. Environmental sustainability of biodiesel in Brazil. Energy Policy. 2014;**65**:680-691. DOI: 10.1016/j.enpol.2013.09.062